Strain and Counterstrain

Lawrence H. Jones, D.O., FAAO

The American Academy of Osteopathy
2630 Airport Road
Colorado Springs, Colorado 80910
1981

© 1981 Lawrence H. Jones, D.O., FAAO
Library of Congress Catalog Card No. 81-67256
ISBN 0-940668-00-9   Hardbound Edition
ISBN 0-940668-01-7   Softbound Edition
All Rights Reserved.

Second printing, December 1983

Contents

Page
5	Acknowledgments
7	Preface
11	Strain and counterstrain
15	An ongoing process
16	Moment of onset
18	Case histories
24	The role of the muscle
26	Modus operandi
28	Tender points
30	Complaints that suggest tender point presence
32	Clues in the search for tender points
33	"Backstrain" not synonymous with dysfunction
34	Self-treatment
35	Prophylactic stretching
37	Assistance needed
38	Glossary of terms
39	Illustrations
109	Index to illustrations
111	References

Acknowledgments

This concept could not have been developed nor this book written without aid from several brilliant men, who advised me, argued with me, instructed me, and encouraged me. Although this list cannot be complete, among those most helpful were: Harry L. Davis, D.O.; Perrin T. Wilson, D.O.; John P. Goodridge, D.O., FAAO; Berkeley Brandt, Jr., D.O.; William L. Johnston, D.O., FAAO; Rollin E. Becker, D.O.; George W. Northup, D.O., FAAO; John H. Harakal, Jr., D.O., FAAO; and Irvin M. Korr, Ph.D.

In addition, I owe a special debt of gratitude for thousands of hours of selfless effort to R. Rex Conyers, D.O., Robert H. Wendorff, D.O., Gerald H. Hooper, D.O., and Harold R. Schwartz, D.O., who, working both separately and with me, assisted in developing a book that could be understood and used by all of our colleagues.

Lastly, I am indebted to those whose suggested revisions added clarity and comprehensibility to the text. This tedious and time-consuming work was done especially by Sara Sutton, D.O., FAAO, and Edna Lay, D.O., FAAO.

Preface

This book has been written in an attempt to pass on to other clinicians the benefits I have been fortunate enough to acquire in 40 years of clinical practice treating joint dysfunctions. It is not a scientific treatise, but the sharing of the experience of one clinician with another.

Especially important and valuable were two phenomena I observed accidentally at different times; each impressed me enough to motivate experimentation with ways to utilize them in treating joint dysfunctions. The first phenomenon provided a unique opportunity to observe the behavior of joints under stress. The second provided a new source of diagnostic information equal to and complementing the knowledge that I had at the time of its discovery.

The hypothesis on the nature of somatic joint dysfunction is derived from an article by Korr,[1] who describes joint dysfunction in a logical manner and provides a scientific basis for understanding it. My observations during some 26 years of using deliberate application of joint strain in treatment bear out the conclusions offered by Korr.

Much of the pain suffered by humans results from one poorly understood dysfunction apparently arising in or around joints. One of the first men in America to put the study of joint dysfunction on a scientific basis was Andrew Taylor Still, a physician living at the time of the American Civil War. He studied the anatomy of the human body diligently, and by reasoning on a physiologic basis he developed manipulative skills, thus greatly improving his capabilities as a physician. Other physicians of his time did not recognize his achievement as beneficial and rejected his ideas concerning the underlying cause of disease. After years of frustration he founded another school of medicine, which he called osteopathy. The objective of the school was "to establish a college of osteopathy, the design of which is to improve our present system of surgery, obstetrics and treatment of diseases generally, and place the same on a more rational and scientific basis, and to impart information to the medical profession. . . ."[2]

Though the new school served to perpetuate his concepts, it further alienated his contemporaries. Still and the graduates of his school, using manipulative methods, were able to relieve patients of conditions where orthodox medical treatment failed. The rivalry that developed served to discredit the efficacy of joint manipulation in the eyes of orthodox physicians until recent years. It is today considered to be a treatment of choice at many large rehabilitation centers.

Practice of osteopathic medicine in the mid-1930s was not standardized. Many thought that joint stiffness and pain resulted from trauma, which caused joint subluxation or a partial or incomplete dislocation. These practitioners applied a specific thrust to effect a release and restore free and comfortable motion to the painful joint. In most cases they were at least partially successful.

To many of us, however, the nature and etiology of this osteopathic lesion were unclear. At best we had only a skill to relieve a painful syndrome. This was the state of the art and science when the author attended a college of osteopathic medicine.

The evolving concept of somatic dysfunction from that of a mechanical type of structure disorder to that of a reflex neuromuscular dysfunction has proceeded irregularly through the years.

The followers of the traditional concept of joint subluxation reduced by sudden, forceful mobilization have obtained fairly good results in most cases over a long period of time.

William Garner Sutherland, D.O.,[3] advanced ideas concerning the function of cerebrospinal fluid and the craniosacral respiratory mechanism, areas not touched by any other of whom I know. One facet that I feel is related to later concepts was his application of slow and very subtle forces applied in the direction indicated by the structures he treated. For example, he observed the movements of the skull in the directions it moved most freely. He found that by this approach he could reduce or eliminate asymmetrical limitations of motion.

Harold Hoover, D.O., developed a method of treating joint disorders, which he called "functional technique."[4] He learned that the anatomic neutral position for these joints developed asymmetrical muscle tensions. Again his direction of movement of the joint was toward that of least resistance and greatest comfort. He envisioned a position of greatest harmony, in which the tensions all around the joint were equal. He called this position "dynamic neutral." Patients in this position developed a progressive lessening of the need for an eccentric position, until his dynamic position and the anatomic neutral position were the same. He was accused of borrowing from the Sutherland concept, which he explicitly denied. I had studied under him in courses offered by the Academy of Applied Osteopathy on the teachings of Harrison Fryette, D.O., and later on his "functional techniques," but I was not yet ready for either method and I returned to the use of low-amplitude, high-velocity thrust-type techniques.

When I published an article in *The D.O.* entitled "Spontaneous release by positioning,"[5] it was suggested that I was borrowing from Hoover's functional technique. I was quick to point out the differences and deprecate the similarities. The position of spontaneous release is in the direction toward immediate ease and comfort — as was his. Whereas his dynamic neutral concept sought a bilateral balance of tension fairly near the anatomic neutral position, the concept of the position of spontaneous release focused on the disorder as unilateral and moved to the position of greatest ease of the abnormally tense side. This was always at or close to an actual position of strain.

One of the earliest pioneers, a man with whom I had the good fortune to discuss concepts, was T.J. Ruddy, D.O. He developed a very effective method of treating joint disorders by a series of rapidly repeated active movements against resistance.[6] He explained the benefits in terms of increased blood flow to affected muscles. He called his method resistive duction. In talking with him it became apparent to me that he thought in terms of malfunctioning nerves and muscles rather than mechanical disorders. He seemed to understand my poorly formed ideas immediately. Unfortunately, he was already of advanced age and I was unable to follow up on our discussions.

Fred L. Mitchell, Sr., D.O., FAAO, the man who developed and taught muscle energy techniques, readily acknowledged Dr. Ruddy as the source of his material.[7] On less self-centered reflection, I am sure that all of these earlier concepts directed my thinking to neuromuscular dysfunction as the basis of joint disorders. What I find heartening and convincing is that on this point, although we each believed that we thought independently, our ideas all converged toward what seems to be a basic truth.

Strain and Counterstrain

This book deals with a new concept of the nature of a common joint dysfunction and with appropriate therapeutic measures to relieve it. Here are offered two definitions:

Strain and counterstrain: Relieving spinal or other joint pain by passively putting the joint into its position of greatest comfort; or *strain and counterstrain:* Relieving pain by reduction and arrest of the continuing inappropriate proprioceptor activity. This is accomplished by markedly shortening the muscle that contains the malfunctioning muscle spindle by applying mild strain to its antagonists. In other words, the inappropriate strain reflex is inhibited by application of a counterstrain.

The above explanations are intended to clarify my beliefs. They are not expected at this point to be accepted as valid. Observed evidence will be presented in detail to show how I arrived at these conclusions.

Probably, the most logical means of presenting these observations is to relate the development of the new concept of the nature of joint dysfunction as it evolved. It began with a most frustrating case of psoitis. Pain and the typical psoitis posture had begun 2½ months previous to my seeing the patient, whose condition had not improved under the care of two chiropractic physicians.

The patient was fit, young and quick, with the grace and power of an athlete, but he couldn't stand erect. He made no progress at all under thrust-type treatment. He complained of night pain and loss of sleep; often he would wake every 15 minutes, try to find a position of comfort, and doze off again. Therefore, I spent one visit finding a comfortable sleeping position for him. It was a tedious process; I would produce a position passively and question the patient as to improvement. After 20 minutes, however, we achieved a position of surprising amount of comfort, the only benefit he had received in 4 months' treatment (one month each by two chiropractors and the balance in traditional osteopathic manipulation by me).

Since this temporary benefit was the only hopeful sign in a frustrating program, I left the patient propped up in his position of comfort for a short while, so that he could observe the position and perhaps return to it that night. But when he stood up he was able to stand erect with comfort. He was overjoyed and so was I, but more than that I was astonished. At that visit nothing had been done for the patient but positioning for comfort, and that had succeeded where my best efforts had failed repeatedly. No doubt, my previous frustration had made this change seem more impressive. It was a primary motivating factor in my experiments with this principle for the 26 years since. The discovery was completely accidental. The new method, if I could learn to apply it to other joint problems, would be far superior to the methods I had used previously.

Experimentation was begun by using this approach in other cases of psoitis. Results were not consistent at first, but the rewards were great enough to encourage perseverance. My first attempts were well tolerated by patients. When I failed, I fell back on older methods. Little by little, results became consistently better, and experimentation with different types of dysfunction began.

This slow approach to comfort and relaxation provided a unique opportunity to observe the behavior of muscles and ligaments under different rates and amounts of stretch. The first observation, one of the most important, was that all benefits could be lost, no matter how successful a position of comfort had been found, if the return from the position to neutral was not done *slowly*, especially through the first few degrees of arc. Beside being essential for successful treatment, this phenomenon later became an important component of a new concept as to the nature of somatic joint dysfunction.

Although many patients did respond just to a position providing the most comfort, it soon became obvious that many cases were more complicated. It is necessary to put all of the joints in one area of the spine into somewhat the same relative position. Some patients had dysfunctions of neighboring joints that required an entirely different position for their relief. They would be aggravated temporarily by the stretch used to relieve the first joint. No lasting aggravation was produced, however, if the return to neutral position was made slowly. The position of comfort method was usually a great help, but it was at times necessary to find evidence of individual joint dysfunctions in a problem area. It was possible to overcome this difficulty by treating the most severe dysfunction first, despite mild pain in a nearby joint, and after that success proceeding to the next most severe.

Of greatest value in diagnosis was the presence in the paravertebral area of small zones of tense, tender, and edematous muscle and fascial tissue. When these were present, the new procedure was indicated. These areas were present and definite in only half of the dysfunctions, however. Other methods of diagnosis, although useful, were not nearly so reliable. The tender points could be probed intermittently during treatment. Success in finding the ideal position of comfort was immediately evident in the form of palpable decrease in tension and in prompt decrease in tenderness of these points.

The means of overcoming this weakness in diagnosis was again discovered by chance, and it was again with a case of psoitis that I found the answer.

I had been treating this patient with fair success, so that he was able to hoe his garden. He struck a rock while hoeing. The deflected hoe handle then struck him in the groin, causing severe pain. Frightened of hernia, he came in for an examination the day before he was scheduled for a treatment. The spot in his groin was exquisitely sensitive, but it hurt only while being palpated. The pain caused by the hoe had not persisted, and, except for tenderness, I found no evidence of hernia. I reassured him and proceeded to give him his next treatment one day early. As I held him in a supine position with his hips flexed and spine rotated for the necessary 90 seconds, I probed the tender spot again. It was not tender. The position of comfort for his psoitis had relieved it, just as a position of comfort had relieved tender spots in the paravertebral area of many back pains.

I wondered if this could be the location of the missing tender, tense areas that could not be found in many of the joint dysfunctions in the paravertebral area. It did not seem possible. The tender spots along the spine were in close relation to the back pain. This tender spot was not associated with any local pain but with back pain. This is the reason that we have been so long in finding these points of tenderness in the anterior part of the body. They are seldom associated with any local pain even though they are often very tender to palpation. The patient, himself, is unaware of their presence until probed by the physician.

Much experimentation did reveal that these anterior tender spots were indeed the sought-after diagnostic points. Further experimentation in cases with inconclusive evidence of joint dysfunction in the paravertebral area revealed the locations of the other 50 percent of the tense, tender points that were so valuable in diagnosis. Every backache lacking definite tender points in the paravertebral area had definite, reliable tender points anteriorly, so I could proceed with confidence in the outcome of any treatment.

By relief of tenderness and reduction in muscle contraction, I was informed reliably of correctness or incorrectness of treatment. I learned to trust implicitly in the information the body provided. For each musculoskeletal or neuromuscular dysfunction there is a tender point, and there is consistently a position that will relieve it.

The location of each of the common myofascial tender points will be presented later with an illustration. Also, an illustration for each of the common positions used will be offered. This is intended to be used only as a guide. The exact location of the ideal position must be determined on an individual basis, and the last degree or two of positioning causes marked changes in tender point response. There has been no effort to present all myofascial tender points nor all positions for relief of joint troubles, since the number of possibilities is limitless. However, guided by just these illustrations and descriptions, the beginner can administer relief to most of the joint dysfunctions encountered in daily practice. He will need to be alert to the possibility of the rare type of dysfunction, but, secure in the knowledge that they are there to be found, he can be prepared for a careful search for a tender point and for the stretch position of release.

An ongoing process

The history and course of conditions termed somatic joint dysfunction differ from those of most ailments to which the body is subject. Response to traumatic injury generally is resolved in one fashion or another. There may be scarring from burns, lacerations, avulsions, or contusions. Even fractures, given time, heal to be sound and pain free. Joint dysfunction, however, has a singularly everlasting, persisting character. Usually it subsides to a relatively quiescent state without treatment, but it rarely leaves completely. Even in its dormant state, a skilled, deep palpation of the affected area will reveal continuing tenderness and tension and resistance when moved in one direction.

There is something wrong — something active — that is ready to flare up on relatively slight provocation. Yet, even after many years of existence, this dysfunction can be stopped by atraumatic positioning. It can be relieved almost immediately, and, by repeated and adequate treatment, it can be maintained in harmonious function, until it heals sufficiently, so that its tendency to recur is inhibited indefinitely.

In patients who suffer from chronic pain, the dysfunction can start mildly, then become progressively worse, and continue to wax and wane in intensity but never to cease. Even these patients can be kept functioning normally until they are healed completely and relieved permanently.

This type of onset and course does not suggest static tissue damage, which would be expected to resolve itself and become pain free within a few months. What seems more likely is that the strain, or the body's reaction to it, initiates an ongoing noxious process, which continues to operate inappropriately so as to serve as an ongoing source of irritation. The body may adapt to the continuing dysfunction so that the pain becomes subclinical for a long period of time, but the body is unable to reverse it.

The philosophy of the physician, then, is aimed not just at promoting healing of a lesion, but at stopping a continuing and irritating dysfunction. He can rely on the body's healing processes to cure the conditions of inflammation, if the normal function can be maintained long enough for the healing processes to complete their work. Indeed, it is apparent that the only condition with which the body is unable to cope is its own neuromusculoskeletal dysfunction.

Moment of onset

If the ongoing distress is a dysfunction rather than tissue damage resulting from trauma, it will be helpful to know the exact time of onset.

Since the title and much of the contents of this book treat of strain, it behooves us to start with definite ideas of what a strain is.

Dorland's Medical Dictionary:[8] "An overstretching or overexertion of some part of the musculature: Sacro-iliac strain: Strain or sprain of the sacro-iliac joint causing constant backache."

Random House Unabridged Dictionary:[9]

Definition 5 — "To stretch beyond proper point or limit."

Definition 22 — "An injury to a muscle, tendon etc. due to excessive tension or use; sprain."

Definition 3 — under Syn. "Strain, sprain imply a wrenching, twisting and stretching of muscles and tendons. To strain is to stretch tightly, make taut, wrench, tear, cause injury to, by long continued or sudden and too violent effort or movement."

Use of the word strain in this book indicates overstretching of muscles, tendons, ligaments, and fascia, with attendant neuromuscular strain reflexes. The focus of attention is directed especially at the neuromuscular reflexes rather than tissue stresses.

In somatic dysfunction we are faced with contradictions between what we view as reality and the continuing reports some part of our body makes. Though we associate the cause of the discomfort with strain, we are unable to understand why the evidence of strain persists when there is no more strain. Pain from overstretching would be expected to stop or be markedly reduced when the strain which we believe to be the cause is completely stopped. Yet, evidence of injury and pain persists, even becomes progressively worse. Objective evidence of the eccentric position in which the body holds the affected joint shows, if anything, an abnormal shortening of the tissues that are reporting strain. For instance, a patient who enters the office bent forward and is unable to stand erect has one or more spinal joints sending continuous messages that they are strained in extension. Any direct force toward extension greatly increases their strain. So our dilemma is reduced to the fact that a strained joint reports a continuing strain and behaves as though it is strained long after actual strain has ceased.

We are aware that this phenomenon does not result from all strains. The response to strain, even if severe, is often what we would expect — a few days of pain and gradual complete recovery. Counterstrain as used in this book indicates a mild strain (overstretching) applied in a direction opposite to that false and continuing message of strain from which the body is suffering.

Some of these problems arise following a severe wrenching trauma with real and obvious injury. These include falls, auto accidents, excessive lifting, or other unusually violent effort. Some result when a strong muscular effort is met with a sudden change in resistance. For instance, a man straining on a wrench is usually in danger only if the wrench slips off the bolt. Or there may be an unexpected increase of resistance, such as if a man scooping grain from the floorbed of a truck strikes a nail projecting from the floor and is stopped abruptly.

More often the physician sees patients with no clear-cut history of violent strain. Their difficulty seems to stem from slight strain, such as bending over. This may be in the range of habitual movement, so that the belief in the related incident as the only trauma must be discounted. The time of onset is often reported as the time of return to a neutral position from a prolonged postural strain rather than the time of the postural strain itself.

Whatever their histories, the problems from which patients suffer and the treatments needed are surprisingly similar. It is interesting to observe little relationship between the amount of trauma reported and the degree of discomfort.

Many of my patients with back injuries were unsure when their pain began. However, those who did recall a specific time usually commented, "It caught when I started to straighten up." Careful questioning, without prompting, elicited a similar response from many other patients. This revelation sets the time of onset for those with a definite opinion *at and during the return from a position of strain*. Acting on this premise, the author experimented with thousands of deliberately applied strains both to patients and his own body. In all tests it was found that even a prolonged strain would not result in neuromuscular joint dysfunction as long as the return from the position of strain to a neutral position was made *slowly* and without force on the part of the strained body. Thus the time of onset of joint dysfunction is not the strain itself, but the body's reaction to strain — a panic reaction and too-rapid attempt to return to the neutral position.

Case histories

The physician is supplied with a wealth of information from his thousands of cases; this knowledge is seldom available to scientists. At least occasionally he observes phenomena that offer insight into the nature of the conditions he treats. The following case histories illustrate, among other factors, the time element related to onset of continuing joint dysfunction. Many case histories offer significant information in the development of a particular concept. A case history that illustrates all of the points of a concept is rare; Case 1 is one such history.

Case 1

The patient, a middle-aged, generally healthy factory worker, had a habit of napping for an hour or so before dinner. Several times during a nap, while lying supine, he would hang his right arm off the edge of the davenport in marked extension. (For a minute or two this position cannot be considered much of a strain, but for a period of 45 minutes, it may become one.) His wife, preparing dinner in the next room, would look in on him occasionally. Worried about his position but not wanting to awaken him, she would very slowly and gently raise his arm and lay it over his chest. This performance was repeated several times over the years, with never a twinge in the man's arm when he wakened for dinner.

One day he napped when his wife was out, and, as before, he kept his arm extended. A phone rang near his head and he awoke with a start, rapidly flexing his overstretched elbow. As he spoke on the phone he became aware of pain in his right biceps. He wasn't a worrier and didn't think much about it, until the pain gradually worsened. There was pain whenever he flexed his elbow, especially against resistance. In time, his biceps actually did become smaller from disuse. Now very concerned, he consulted physicians, who on roentgenographic examination diagnosed strain of the biceps. Surgery was considered but not performed because careful probing did not demonstrate tenderness in the biceps.

The man had been suffering for 2 years when he consulted me. Like the other physicians, I failed to elicit tenderness in the biceps. However, palpation revealed sharp tenderness beside the olecranon process of the ulna. (This is evidence of excessive proprioceptor activity in the triceps.) The triceps had never been overstretched, but had suffered only prolonged over-shortening and sudden, panic-type lengthening. Yet, despite location of pain in the biceps, the abnormal myofascial tender point was situated in the triceps.

Treatment consisted of positioning the elbow in hyperextension so that the biceps was put on a stretch and the triceps was allowed to shorten maximally. If there had been tissue rupture in the biceps, surely this stretch would have aggravated the patient's condition. He was relieved immediately and left the office with half his pain gone. A few more similar treatments brought complete recovery and there was no recurrence.

The case is unusual only in that it demonstrates the importance of slow return from a strained position. The patient's wife did not know this; she was moving his arm slowly to avoid waking him. Yet her actions had served the purpose of aborting a potential joint dysfunction on several occasions. His pain did not start as long as his wife slowly returned his hyperextended elbow to a neutral position. The pain began because of his sudden flexion of a joint that had been overstretched in extension. Even more surprising, the continuing dysfunction, complete with its tender point, was on the posterior aspect of the elbow, where tissues had never been strained. The real strain, the pain, and the apparent weakness were on the front of the elbow and biceps, but the tender point was on the extensor side of the elbow related to the triceps. Atrophy of the biceps resulted from disuse; it recovered completely.

Case 2

A young man was running down some steps with low risers and wide treads, such as are found at the entrance of many public buildings. He ran a little too fast and overstepped with his left foot, so that only his heel landed on the edge of the step. His ankle was sharply and painfully strained or sprained in extension. There was residual pain and a weakness in lifting his foot. He scuffed his toes and occasionally stumbled, and he learned to raise the left knee higher than the right. The affected foot was strong enough for him to stand well, and he could even hop on it.

An orthopedist advised surgery for torn ligaments. The patient, not wishing to bear the substantial costs surgery would incur, consulted me. My examination revealed no tenderness in the anterior ankle. Sharp tenderness was found by palpation beside the attachment of the Achilles tendon. Treatment incorporated positioning of the ankle in hyperextension, which allowed hypershortening of the gastrocnemius muscle, holding the position for 90 seconds, and slowly returning to the neutral position. Two additional treatments were sufficient for lasting comfort. The tender point and evidence of excessive proprioceptor activity were again on the opposite side from the real strain, the pain, and apparent weakness.

Case 3

A middle-aged businessman arrived home a little early for lunch one spring morning, so he squatted down and pulled a few weeds in the flower garden. He was still at it 45 minutes later. Finally, his wife called him, and he arose suddenly and felt a low-back pain, which prevented him from standing erect. Examination revealed a myofascial tender point in the psoas muscle. Treatment included marked thoracolumbar flexion until the tenderness subsided, holding the position for 90 seconds, then slowly returning to normal. Although the pain was in the back, there was no posterior tender point. Assuming that 45 minutes of squatting constitutes a strain for a businessman, the strain surely must have been in the posterior part of the spine. Apparently the only injury the psoas muscle had suffered was prolonged shortening and sudden lengthening. Just one treatment was needed because the patient sought treatment promptly.

Case 4

A lady of 70 years sometimes would nod off, then awaken with a start and raise her head off her chest. After one particularly long doze with her head down, followed by an abrupt awakening, she was unable to hold her head up, because of severe lower cervical pain radiating across the top of her shoulders. No tenderness was found posteriorly where the pain was located, but there was sharp sensitivity in the suprasternal notch, indicating dysfunction at the level of the first thoracic intervertebral joint. With the patient seated, and with her hands clasped over the top of her head, it was easy to attain marked cervicothoracic flexion at that level. Ninety seconds of holding this position and slow return to neutral relieved her pain. She continued to have recurrences for several months. I assumed that this was because she was repeating the original strain.

Far from being rare, this type of scenario is so standard as to become almost predictable. In many cases I examine the area near the pain just because a patient expects it. He is impressed at my finding a sharp tender point where he has no pain, but he would be disappointed if I didn't examine the site of the pain.

Except for the ankle case, these histories follow a similar course — that of prolonged strain and sudden recovery. Regardless of the type of strain, whether prolonged or sudden and violent, successful treatment follows the same pattern.

Conclusions and summary

Since attaining the position of successful treatment is simultaneous with confirmed diagnosis, we may assume that the type of trauma is not important.

We now have had an opportunity to observe facts about the behavior of the joint dysfunction in motion which, taken together, may enable us to arrive at some conclusions. Perhaps they will suggest the nature of and the cause of joint dysfunction.

1. The pain of joint dysfunction is position oriented, from severe pain in one position to almost complete comfort in the exactly opposite position.

2. Dysfunction does not result from the strain itself, as had been supposed, but from an occurrence during the time the body was reacting to the strain.

3. Palpable objective evidence of the continuing dysfunction is found not in the tissues that were overstretched, but in the tissues that are the antagonists of the overstretched tissues. The history of these tissues seems essentially atraumatic. They suffered only extreme shortening followed by a panic-type reaction involving very rapid lengthening.

4. Evidence obtained by treating many thousands of these dysfunctions in a position of comfort verifies that to obtain success in treatment, the rate of return to the neutral position must be *slow*. Otherwise, the return to a neutral position is likely to precipitate a return of the dysfunction.

5. The behavior of this dysfunction seems entirely related to joint

position, so we are brought to the realization that the joint is reacting as if it were actually strained; however, it is not strained but slightly shorter on the affected side than on the contralateral side. If a normal, healthy joint were put into a position of strain, increasing the position of strain would aggravate the pain and moving out of that position would ease it. The joint dysfunction acts then as if it were continuously being strained.

6. The structures of the body responsible for reporting the positional states of joints are the proprioceptor nerve endings. Everything we have learned about joint dysfunction points to them as the major site of beginning disorder. What could go wrong with a proprioceptor nerve ending? In 1975, Korr[1] advanced the concept of a neural basis for joint dysfunctions, incriminating in particular the muscle spindle or primary proprioceptor nerve endings. He stated, "One distinction may be worth mentioning. Although discharges of both types of endings are more or less proportional to length, the primary (annulospiral) ending has the additional feature that its frequency of firing *during* a stretch is in proportion to the *rate* of change. That is, the secondary ending apparently reports length at any moment, but the primary ending reports both velocity of stretch (and hence of joint motion) and length (hence joint position). The primary ending, thereby, provides a predictive or anticipatory input to the nervous system."

Now let us visualize the primary ending in the muscle spindle of the muscle on the opposite side of the joint from the side being strained. It is at the minimal limit of its stretch, extremely short. The input from its proprioceptors is probably almost nil. Add to this reciprocal inhibition from the reflexes on the overstretched side and visualize an extremely low rate of firing of impulses from these proprioceptors. Now the body reacts to the emergency of the strain and suddenly and forcibly straightens the joint, stretching the hypershortened muscle and its proprioceptors so that *it begins to report strain even before it reaches its normal length*. Once begun, this inappropriate message of strain when there is none cannot be turned off by the body.

One would think that the central nervous system, through a reduced outflow of the gamma motor neuron, would relax intrafusal fibers enough to restore the primary proprioceptor to a normal rate of firing. Korr[1] offers the hypothesis that the central nervous system, seeking a response from the hypershortened and silent primary ending, begins an extraordinary outflow, which, followed by an unusually fast stretching, results in high gamma gain that the body is unable to reduce to normal.

What is accomplished in treatment by positioning for comfort? The position of comfort is identical with that of the original strain. This position again shortens the muscle containing the dysfunctioning proprioceptors. Despite their continuing dysfunction, this serves to allow the primary and secondary endings to cease their abnormal activity. Korr[1] suggests: "The shortened spindle nevertheless continues to fire, despite the slackening of the main muscle, and the CNS is gradually enabled to turn down the gamma discharge, and, in turn, enables the muscle to return to 'easy neutral' at its resting length. In effect, the physician has led the patient through a repetition of the lesioning process, with, however, two essential differences: first, it is done in slow motion with gentle muscular forces, and, second, there have been no 'surprises' for the CNS; the spindle has continued to report throughout."

As with counterstrain technique, the original stretch must be maintained for 90 seconds to allow for adequate decrease in firing from proprioceptors. To avoid re-exciting the dysfunction of this highly facilitated reflex, the structures must be returned to their neutral position very slowly. During the 90 seconds, the palpating finger observes further subtle tissue changes, including decrease of "bogginess" and further softening (lessening of tension).

The role of the muscle

Examination of tissues in the paravertebral area often reveals the presence of tense muscles in the area of discomfort. Because joint dysfunction has been thought of as any kind of mechanical restriction of motion, that is, subluxation, one would probably imagine the muscle to be striving vainly to overcome the supposed articular restriction of motion. However, the newer concept of dysfunction and its treatment visualizes the opposite.

Figure 1 is a 3-part schematic representation of a joint and its muscles. Figure 1-I depicts the joint at rest. Primary nerve endings are sending the usual rate of flow of proprioceptor impulses into the central nervous system to indicate a tonic condition of the muscles. The tone of one muscle is balanced with the other.

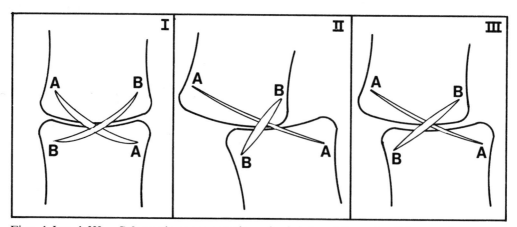

Figs. 1-I to 1-III. **Schematic representation of a joint and the role of its muscles.**

Figure 1-II illustrates a condition of severe overstretching of muscle A. It is initiating an increased frequency of impulses from both primary and secondary endings. On the other hand, muscle B is exceedingly unstressed and hypershortened, so that it is firing a minimum of impulses into the central nervous system. This is a very unpleasant situation and the body reacts to it. If it is able to react with some deliberation, there will have been an overstretching, but nothing more. This joint can be returned directly toward neutral without pain if it is done slowly. If, however, as often happens, this joint response is highly emergency oriented and there is a panic-type, all-out effort to recover from the painful overstretching, a sharp reflex muscle contraction will result. Surprisingly, this occurs not in the overstretched muscle A, but in the unstrained antagonist muscle B. The direction of the ensuing deformity is such that the joint is not able to return completely from its overstretched position.

This is the position demonstrated in Figure 1-III, a position not nearly as abnormal as the one in Figure 1-II, yet one not able to be straightened directly toward a neutral position. It can be seen that muscle B cannot relax adequately. If we were to try to force this joint to return to the position in Figure 1-I, we would encounter increasing resistance to the point of pain and rigidity. As soon as we would release the involved parts, the joint would return to the position depicted in Figure 1-III. There would definitely be joint dysfunction. No matter how slowly we would try to return it to the neutral position directly, we would fail. However, if we reposition it as it was in Figure 1-II, it doesn't hurt but feels better and seems to relax. Then, after 90 seconds, if we slowly return it to the neutral position, we succeed. Muscle B has relaxed and lengthened and allows the joint to remain neutral. The muscle that was overstretched might be slightly sore for a few days. The sudden stretching of muscle B caused the muscle spindle to report to the central nervous system that it was being strained even before it reached normal length. The pain in muscle A is simply a manifestation of pain produced by stretching the resistant muscle.

Modus operandi

Finding the myofascial tender point and the correct position of release will probably take a few minutes at first. Watching a skilled physician find a tender point in a few seconds and the position of release in a few seconds more may give a false impression of simplicity to the neophyte. The first experience for the author took 20 minutes, and that was just searching for a position of comfort. The novice may expect to make several false moves before he makes the right one. And it might take a long time to locate the ideal position. For example, assume that he has located a tense, tender spot by deep palpation. The pressure he used for best results was just short of the amount to cause pain in normal tissue, so that a hypersensitive spot was apparent both to the patient and the doctor. Assume that the physician is unaware of the suggested direction of movement for positioning and actually moves his patient the wrong way. In Figure 2, the midline indicates a neutral

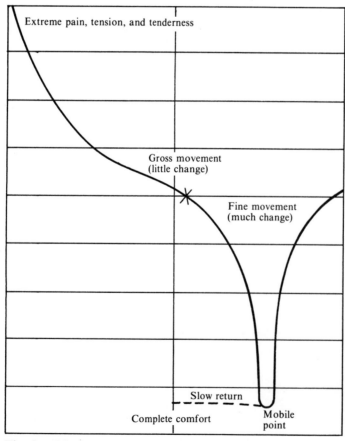

Fig. 2. **Modus operandi for ideal positioning for comfort.**

position, and X indicates the position that the joint had assumed when the tender point was found. If the joint is positioned toward the left, the muscle tension increases. At some point before he reaches the area of extreme tenderness, he becomes aware that this is the wrong position. The amount of muscle tension and tenderness near the midline varies moderately with movement, and the novice has likely spent previous efforts palpating static conditions, not changing conditions. He tries the opposite direction and begins to feel that he is having some success. The patient reports a little less tenderness, and the area seems slightly less tense to the physician, so he continues. As he arrives eventually within a few degrees of his target position, tissue changes are considerable, and with only slight movement. At first, he moved through a large arc; now he must move more and more slowly or he might pass through the target position without recognizing it. He may feel an abrupt, marked relaxation of the tension, with tension recurring again if he moves another degree or so. He should return to the position of abrupt relaxation; this is his goal. If he moves from this position he may feel tension begin abruptly in just about any direction, in as little as one degree of movement. This point of maximum relaxation accompanied by an abrupt increase in joint mobility within a very small arc is the "mobile point." This term will be used to designate the ideal position of comfort. After holding this position for 90 seconds, the physician slowly moves the patient back toward the neutral position, with only slight increases in tenderness and tension. He has made his first correction of a joint dysfunction by use of counterstrain techniques.

    The increasingly slow approach through a smaller and smaller arc which is done at or near the ideal position has been termed "fine tuning."

Tender points

These tense, tender, and edematous spots on the body have been known for countless years in the Orient, along with other, slightly different but often closely related points, such as acupuncture sites. The points I have found most useful were first termed "myofascial trigger points" by Travell, who treated them with injection or with a coolant spray to effect an improvement.[10] Travell named them trigger points because stimulation produced referred pain. Similar palpable points were described by Chapman, who related them to visceral function.[11]

The tender points used in counterstrain techniques are not located in or just beneath the skin as are many acupuncture points, but deeper in muscle, tendon, ligament, or fascia. They measure 1 cm. across or less, with the most acute point about 3 mm. in diameter. They may be multiple for one specific joint dysfunction, may extend for a few centimeters along a muscle, or may be arranged in a chain (such as the ones in the muscle and fascia along the lateral surface of the femur). A physician skilled in palpation techniques will perceive tenseness and/or edema as well as tenderness, although the tenderness, often a few times greater than that for normal tissue, is for the beginner the most valuable diagnostic sign. He maintains his palpating finger over the tender point to monitor expected changes in tenderness. With the other hand he positions the patient into a posture of comfort and relaxation. He may proceed successfully just by questioning the patient as he probes intermittently while moving toward the position. If he is correct, the patient can report diminishing tenderness in the tender area. By intermittent deep palpation he monitors the tender point, seeking the ideal position at which there is at least two-thirds reduction in tenderness.

Tender points found in the paravertebral area are often closely related to an area of pain. Those found in the anterior part of the body usually are free of pain at the site of the tender point until it is probed.

The anterior tender points are associated with manifestation of spinal joint dysfunction and pain in the posterior part of the body. Relatively few patients complain of pain in the anterior torso, even though anterior tender points are present. The tenderness elicited in the anterior aspect is often surprisingly acute to the patient, who is usually unaware of its presence until pressure is applied. Likewise, tender points in appendicular areas, such as the extremities, are often found in a painless area opposite to the site of pain and apparent weakness.

Naming of anterior tender points was done after considerable attempts were made to correlate them with specific segments of the spine and pelvis. Though admittedly somewhat arbitrary, they have proved to be accurate enough to serve well for effective diagnosis and treatment. Charts of the most common tender points along with illustrations of recommended treatments are presented later in the book. These provide general information for the majority of problems encountered in daily practice. No attempt is made to provide all possible tender points or treatments; the number of possibilities is limitless. Rule-of-thumb suggestions for location of the rarer tender sites plus intuitive skill will provide the earnest worker a broad scope of capability.

The question is asked whether the repeated probing of the tender point is therapeutic, as in acupressure or Rolfing techniques. It is not intentionally therapeutic, but is used solely for diagnosis and evidence of accuracy of treatment.

There is a tendency among some practitioners to think of the tender point itself as the disease entity. However, it is only a palpable manifestation of a joint dysfunction. Counterstrain treatment, oriented exclusively toward release by positioning, is successful in treating the joint dysfunction. The tender point ceases to report tenderness when normal function is restored to the connective tissue related to the joint.

The majority of somatic dysfunction cases that the physician will encounter in his office can be classified under a few dozen types of joint problems. These can be presented along with probable gross solutions to bring each specific treatment into the realm of the ideal position. If the physician understands the principles, he will be able to discover the tender point and the position of treatment for any of the unusual problems presented by patients.

Complaints that suggest tender point presence

This outline of common complaints with related tender points is offered as an aid in diagnosis:

Headache
        Frontal, in or behind the eye or with photophobia — First cervical or occipitomastoid areas
        Periorbital — Second cervical level; occipitomastoid, squamosal, infraorbital, and nasal areas
        Occipital area to vertex — Fourth cervical level
        Vague, generalized — Fifth cervical level

Earache — Third cervical, posterior auricular, zygomatic, masseter, and occipitomastoid areas

Tinnitus — Third cervical, posterior auricular, zygomatic, and occipitomastoid areas

Vertigo — Third cervical and occipitomastoid areas

Maxillary sinus — Infraorbital and second cervical areas

Dental neuritis
        Upper — Squamosal, sphenoid, and lateral canthus areas
        Lower — Masseter area

Neck pain — Tender points may be anterior

Dysphagia — Occipitomastoid, other cranial areas, trachea or hyoid bone, inion, and anterior first cervical level

Cough (nonproductive, throat tickle) — Anterior fifth, sixth, or seventh cervical levels found on the trachea

Precordial pain — Anterior third, fourth, fifth, or sixth thoracic ribs or interspaces

Heartburn — Anterior fifth thoracic level

Fatigue, especially on awakening — Fifth, sixth, or seventh thoracic areas, especially anteriorly

Epigastric pain, gastritis, ulcer, diaphragmatic hernia — Anterior seventh or eighth thoracic area (effectively augments medical treatment)

Umbilical pain — Anterior ninth or tenth thoracic area

Diarrhea or constipation — Anterior ninth or tenth thoracic area

Cystitis (noninfectious) — Anterior eleventh thoracic and fourth lumbar areas, especially anteriorly

Shoulder pain
        Steady — Upper four thoracic levels and ribs

With arm movement — Upper four thoracic levels and ribs; in addition, dysfunction of shoulder joint

Pain in entire arm — Second, third, or fourth thoracic and rib areas; sixth, seventh, or eighth cervical level

Pain, numbness in hand or fingers — Sixth, seventh, or eighth cervical and first thoracic levels

Elbow pain

    Without stiffness — Tender point high on the lateral epicondyle, first thoracic or rib level; tender point high on medial epicondyle, fourth thoracic or rib level.

    With stiffness on pronation — Radial head and tip of epicondyle of humerus (epicondylitis)

    On extension — Coronoid areas

    On flexion — Olecranon areas

Wrist pain (local) — Local area, also check radial head

Carpal tunnel syndrome — Many are simply flexed wrist dysfunctions

Thumb pain and weakness — First carpometacarpal area and radial wrist flexion; also check radial head

Pain and weakness in grasp — Dorsal carpometacarpal area

Groin pain — Anterior twelfth thoracic and first lumbar joints and anterior tender points for hip enarthrosis, especially anterior medial trochanter and inguinal ligament

Low-back, hip, and thigh pain — Lower three thoracic, all lumbar, all sacroiliac, and enarthrosis areas (the latter often have a marked positional influence)

Coccygodynia — A large proportion of these conditions is related to a sacroiliac dysfunction, with ilium high in back and flared out above

Clues in the search for tender points

Several rules of thumb will aid in the search for tender points. Many patients are unaware of any position of comfort, although most know of a position or effort that causes pain (the opposite of the position of force applied for comfort).

Any abnormal posture of the body probably will be corrected by exaggeration of the abnormality of position.

Weakness or pain on exertion usually occurs when an effort is made against any resistance (usually gravity) that stretches the muscle containing the tender point, although weakness usually is felt in its antagonist.

Most spinal pains are felt in the posterior part of the body, even though half of the tender points for spinal pain are found anteriorly.

Anterior tender points are found in dysfunctions eased by flexion (forward bending). Posterior tender points are found in dysfunctions eased by extension (backward bending).

The nearer to the midline (the central sagittal line) that the tender points are found, the more extension or flexion will be needed in relation to sidebending. The further lateral to the midline that the tender point is found, the relatively greater amount of sidebending (usually away from the painful, tender side) will be needed for relief in relation to extension or flexion. The bending away from the area containing the tender point might be explained by the mechanics of an unstable vertebral spinal column arranged vertically and acted upon by gravity. The side of the lateral spinal convexity does the work of resisting gravity. It is relieved by exaggeration of the existing lateral convexity, which apparently stretches the posterior spinal muscles containing the tender points.

"Backstrain" not synonymous with dysfunction

Physicians have listened to histories of "backstrain" and back pain so often that they may automatically think of them as synonymous. However, strain is a one-time tissue injury resulting from overstretching. Healing should be complete within a few days. Sprain differs only in degree, with definite rupture of tissues and greater inflammation than injury described as strain. Neither should give rise to the long-lasting and often progressive pain syndromes that the physician encounters in patients with continuing joint dysfunctions. A strain or sprain is an injury with which the reparative processes can cope, while histories of continuing dysfunctions demonstrate that they are problems with which the reparative processes cannot cope. There is a popular saying among laymen that "a strain is worse than a break." Probably the reason for this attitude is that pain from the "strain" which becomes a joint dysfunction is everlasting, whereas the fracture eventually heals and becomes pain free.

Self-treatment

Most physicians I have known have accumulated a fund of joint dysfunctions in their own bodies; this irony may be explained by several facts. The treatment with thrust to cause joints to "move" relies in part upon the element of surprise. The patient may have a fairly clear idea of what will happen to him, but he will not be aware of the exact moment of thrust. This advantage to the physician is lost in any attempt at self-treatment. It is impossible to surprise oneself. So why don't we have our colleagues care for us? Usually, we don't want to bother them except in case of dire need. Through the years, physicians, like their patients, accumulate joint dysfunctions in several parts of their bodies, both clinically and, perhaps, especially subclinically. For treating oneself, a search for a position of stretch that releases muscle tension around a joint and restores normal motion, requiring only a slow return from the stretch, is much more effective. True, treatment often will be different from that used on patients, but the principles of applying a specific stretch to a particular joint should be adhered to. A second book, probably designed for lay readers, will present instructions in detail.

Prophylactic stretching

Most physicians who practice for a long time in one area collect their quota of "old crocks," patients with unending arrays of ailments which vary from time to time but never cease. It is easy to label these patients as hypochondriacs or as having similar psychological disorders, but many seem to have real problems with objective evidence. The physician may feel that the unending list of complaints stems from a number of subclinical dysfunctions lying just below the surface, ready to erupt on slight provocation.

The overworked physician dreams of a nirvana where all of these minor ailments could be eliminated completely with recurrence unlikely. I hoped to devise some method of restoring suppleness and comfort to the entire body by use of the principles described previously.

The first concept — to put each joint of the body into all possible positions, maintain it for 90 seconds, and return it slowly to neutral — would be impractical, if only because of the time required. It would be necessary to devise a method that would affect large numbers of joints at one time. Possibly the entire spine could be stretched at one time, I thought; this too was impractical timewise. Finally, an attempt was made to stretch the whole spine in four directions and hold each position for 90 seconds. This did not often achieve an ideal position of comfort. But, by subjecting all of the spinal joints to all of the major stretches, the body would be exposed piecemeal, so to speak, to one element at a time. Therefore, perhaps all of the elements of any joint dysfunction could be satisfied part by part. So I began experimentation by stretching my own spine forward, backward, and to either side, and by rotating to either side.

Stretches were devised so that there would be no necessity to return to neutral from any stretch to overcome the force of gravity. Thus, side stretches were made in the supine position. Rotations were made with the shoulders supine and the pelvis rotated by the legs. Forward and backward stretches were performed with the body lying on either side. Stretches were initiated and maintained by use of all body muscles for approximately 90 seconds, and then the muscles were slowly permitted to relax. With no force of gravity this accomplished partial straightening, and, after a short wait, the rest of the return was done slowly. During my first attempts, I discovered some joint limitations of which I had been unaware. Although they hadn't hurt enough to reach my threshold of consciousness in ordinary positions, they were brought to my awareness by a marked stretch. I prudently avoided any joint limitation that was evidenced by a moderate amount of pain during the first few attempts, thinking I might cause the development of new dysfunctions. Later these fears proved unfounded, provided that I adhered to the rule of slow return.

The immediate subjective response was a feeling of relaxation and suppleness, but I have always distrusted subjective responses. The next morning I developed a soreness and stiffness, which I attributed to the same phenomenon I warn all my patients about at the first visit — the likelihood of a reaction to changes in function of treated joints. They are often quite sore for a few days following even a treatment as nontraumatic as counterstrain. There was no apparent reaction of soreness following succeeding stretches, however. About this time it was easy to convince myself that I had an increased sense of well-being and that I slept better. Later, other joints were brought into the program; the sacroiliac joints were stretched against each other and the hip and shoulder joints were stretched in various ways. After a few weeks of this program (performed twice a week), I could feel no further improvement. I seemed to be maintaining a high level of well-being. When I stopped for a few months, vague disorders appeared, each one mild and unimportant, and I also noticed more weariness after work and less restful sleep. A second course of treatment relieved these problems, and after 3 weeks I stopped again. I would probably be better off if I stretched continuously, but I feel capable of restoring high-level well-being whenever I wish.

The next step was to encourage patients to use this method. This attempt met with mixed success. Patients often preferred to receive their treatments instead of trying to help themselves. Others, however, because they lacked funds or because they were more oriented to self-maintenance of health, accomplished much improvement and needed far fewer office calls.

This method seems to be effective and safe in a large proportion of cases if it is done according to instructions. It should be done at a time when it can be unhurried and undisturbed.

Assistance needed

There is much more to be learned about the use of counterstrain. As more physicians adopt this approach to treatment, they will discover disorders not listed here or better ways to succeed with common dysfunctions. The author will be grateful for any information received from practitioners of this method. If a later edition of this book is published, their new knowledge will be included. Please contact the author or the office of the American Academy of Osteopathy.

Glossary of terms

Anterior refers to a tender point found on the anterior half of the body. With the possible exception of the anterior fourth cervical vertebra, it indicates a joint dysfunction that will require some forward bending in its treatment.

Posterior refers to a tender point on the posterior surface of the body and indicates a joint dysfunction that will require some backward bending in its treatment. Exceptions are the inion, third cervical vertebra, lower pole fifth lumbar vertebra, and piriformis. These are four dysfunctions with posterior tender points that are treated with some forward bending.

Rotation right refers to spinal rotation whereby the body of the upper of two vertebrae is rotated toward the right in relation to the body of the vertebra below. This is confusing, particularly when moving the lower vertebra, as, for instance, when moving the legs to influence the spine. Whichever way the lower vertebra rotates in relation to the one above, the accepted definition of rotation would be the reverse. For example, if the upper of two vertebrae is turned to the right, the lower is turned to the left in relation to the one above. Then, if the legs are rotated to the right, the classic description would be left rotation.

High ilium, low ilium, and flare out refer to abnormalities of the position of the posterior surface of the ilium in relation to the sacrum.

Ankle and foot motions: Ankle flexion approximates toes and knee; ankle extension tends to point the toes, that is, separates the toes and knee. Foot flexion sometimes is called plantar flexion; foot extension sometimes is called dorsiflexion.

The application of forces is directed toward producing the most favorable bend and rotation for the particular segment under treatment. (For instance, to treat a depressed fifth rib on the right, the patient is seated leaning toward the left and is supported so that the fifth segment of his spine is sidebent right and rotated right.)

Illustrations

The photos and drawings aid in finding the approximate positions for relieving specific somatic joint dysfunctions. They must be modified slightly in most cases to locate the ideal position of release. Positional changes are made slowly through small degrees of arc and are guided by tissue relief of tenderness and tension. Remember to expect marked tissue changes in response to relatively fine positional adjustments as the ideal position is approached. A common error is to move completely through the ideal position so rapidly as to fail to recognize it.

In case it should not be apparent, illustrations are for dysfunctions found on the right side of the body. However, for clarity, some of the tender point locations shown in the diagram on page 40 are placed on the left side. Where it was thought helpful, arrows were added to indicate the direction of the force applied. The numbers indicate the estimated amount of force in kilograms. Do not let the concentration on hands distract you from the real objective — the amount, kind, and direction of stress applied to the specific joint being treated.

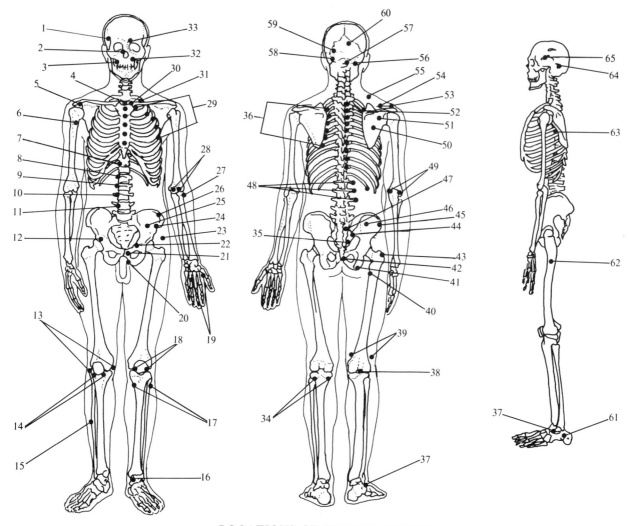

LOCATIONS OF TENDER POINTS

1. Squamosal
2. Nasal
3. Masseter-temporomandibular
4. Anterior first thoracic
5. Anterior acromioclavicular
6. Latissimus dorsi
7. Anterior seventh thoracic
8. Anterior eighth thoracic
9. Anterior ninth thoracic
10. Anterior tenth thoracic
11. Anterior eleventh thoracic
12. Anterior second lumbar
13. Medial and lateral meniscus
14. Medial and lateral extension meniscus
15. Tibialis anticus medial ankle
16. Flexion ankle
17. Medial and lateral hamstrings
18. Medial and lateral patella
19. Thumb and fingers
20. Low-ilium flare-out
21. Anterior fifth lumbar
22. Low ilium
23. Anterior lateral trochanter
24. Anterior first lumbar
25. Iliacus
26. Anterior twelfth thoracic
27. Radial head
28. Medial and lateral coronoid
29. Depressed upper ribs
30. Anterior eighth cervical
31. Anterior seventh cervical
32. Infraorbital nerve
33. Supraorbital nerve
34. Extension ankle (on gastrocnemius)
35. High flare-out sacroiliac
36. Elevated upper ribs (on rib angles)
37. Lateral ankle
38. Posterior cruciate ligament
39. Anterior cruciate ligament
40. Posterior medial trochanter
41. Also posterior medial trochanter
42. Coccyx (for high flare-out sacroiliac)
43. Posterior lateral trochanter
44. Lower-pole fifth lumbar
45. Fourth lumbar
46. Third lumbar
47. Upper-pole fifth lumbar
48. Upper lumbars
49. Medial and lateral olecranon
50. Third thoracic shoulder
51. Lateral second thoracic shoulder
52. Medial second thoracic shoulder
53. Posterior acromioclavicular
54. Supraspinatus
55. Elevated first rib
56. Posterior first cervical
57. Inion
58. Left occipitomastoid
59. Sphenobasilar
60. Right lambdoid
61. Lateral calcaneus
62. Lateral trochanter
63. Subscapularis
64. Posteroauricular
65. Squamosal

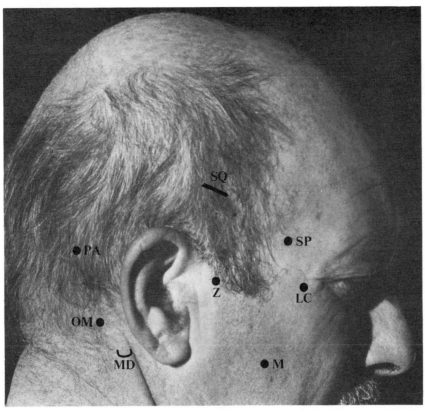

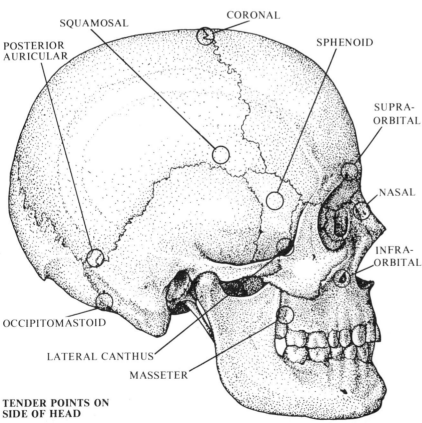

TENDER POINTS ON SIDE OF HEAD

Fig. 3. **Tender points on the side of the head.** The one on the squamosal suture (SQ) is best probed from above. The posterior auricular point (PA) is located in or on the margin of a depression in the skull about 4 cm. back from the pinna near the top. The occipitomastoid tender point (OM) is found in a vertical depression just medial to the mastoid process of the temporal bone, about 3 cm. above the mastoid tip (MD). The sphenoid point (SP) is over the greater wing of the sphenoid bone in the temple. Located about 2 cm. back from the lateral canthus of the eye is the lateral canthus tender point (LC). The zygomatic point (Z) is just above the zygomatic process of the temporal bone, and the masseter point (M) is located on the anterior border of the ascending ramus of the mandible. Observe that these cranial tender points are named from their location on the skull and not from the joint to which they relate. Cranial studies followed use of these tender areas for all other parts of the body, but the sutures of the skull need a lifetime of study. (To me this situation is comparable to playing chess with 100 men on a side, planning 5 moves in advance. By the time I had begun to adapt my method to treat cranial disorders, I had acquired an abiding faith in the reliability of the tender points to report the efficacy of treatment. I claim no mechanical understanding of the skull, but I am able to relieve most cranial problems simply by relying on feedback from the tender points. This method probably is not comparable to the cranial studies developed by Dr. Wm. G. Sutherland and used by skilled practitioners. It is much easier to learn, however, and it does an excellent job. On these terms I am willing to forego mechanical understanding.)

Fig. 4. **Posterior cranial tender points.** This photo attempts to indicate lambdoidal sutures. I indicates landmark of the inion, and MD, another landmark, the tip of the mastoid process. Tender points include the lambdoid (L), just medial to the lambdoid suture and 2-3 cm. down from the lambda itself. The sphenobasilar tender point (SB) also is just medial to the lambdoid suture and obliquely up and lateral to the inion. The occipitomastoid (OM) point is located in a shallow, vertically directed depression about 3 cm. above and slightly medial to the tip of the mastoid process. A first cervical tender point (1C, inion) is found on the medial side of the main posterior cervical muscle mass where it attaches to the occiput about 3 cm. below the inion. It is one of three first cervical tender points treated as an anterior dysfunction (See Fig. 23).

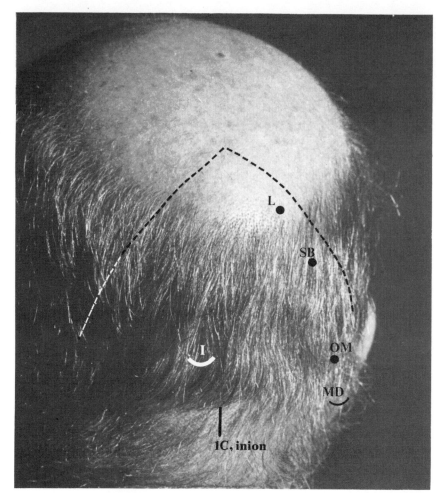

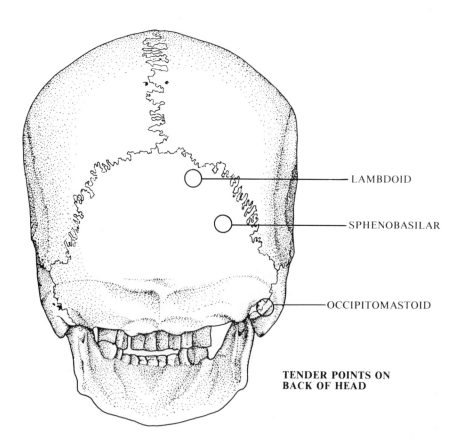

TENDER POINTS ON BACK OF HEAD

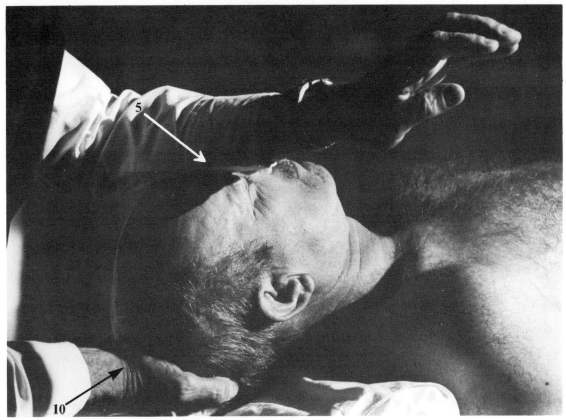

Fig. 5. **Usual treatment for occipitomastoid problems.** The "Y" force is applied with the heel of the right hand caudad and forward obliquely near the top of the occiput. The left forearm applies force caudad and backward over the frontal bone. The lower hand must exert more force. There is a tendency to bend the head over the top of the neck over a transverse axis. It is more comfortable to the physician if his right elbow is tucked into his abdomen so that body force does the work of the right shoulder.

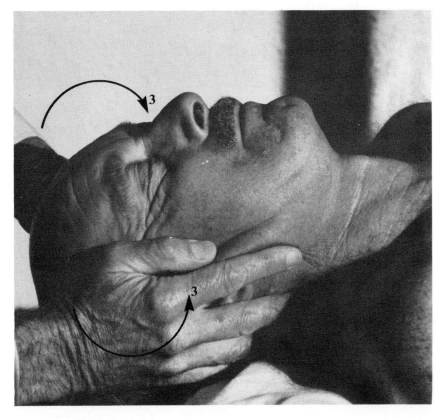

Fig. 6. **A less common treatment for occipitomastoid disorders.** If the problem is unilateral, as usual, it may be better relieved by this "unscrewing" force on the head. Start with medial compression applied over the sides of the back of the head with the palms. One side is twisted as if to unscrew a jar lid, and counterrotation is applied to the opposite side. This is a variable condition which is sometimes relieved with rotation in one direction and sometimes in another.

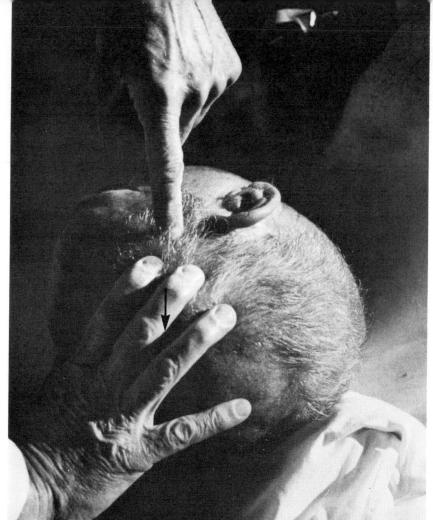

Fig. 7. **Treatment for squamosal dysfunction.** This is a common, important cranial disorder. If probed from above, the border of the squamosal portion of the temporal bone may be palpated. Release is obtained by the pressure of three fingers on the lower part of the parietal bone pulling away from the temporal bone. Uncommonly, this tender point is further forward on the squamosal suture, and the pull is made over the frontal bone obliquely forward. In both this case and for the posterior auricular dysfunction shown in Figure 8, the site of placement of the hand on the skull is that that most effectively moves the tissues of the tender point over the skull. Also, a pillow should be placed under the other ear; this treatment very likely will give relief to upper dental neuritis.

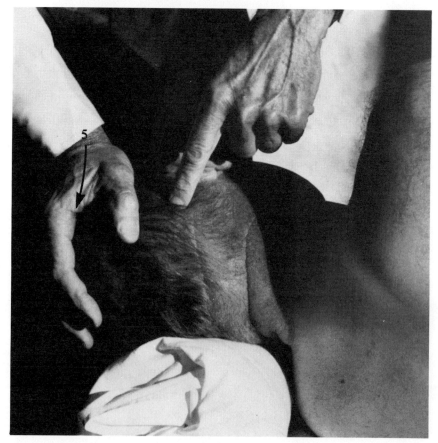

Fig. 8. **Treatment for posterior auricular dysfunctions.** The patient lies on the left side, and a small, firm pillow is rolled up under the left zygoma and ear. Five kilograms of pressure from the heel of the right hand is applied downward so as to bend the skull sideward over an anteroposterior axis. This tender point may be related to tinnitus or deafness.

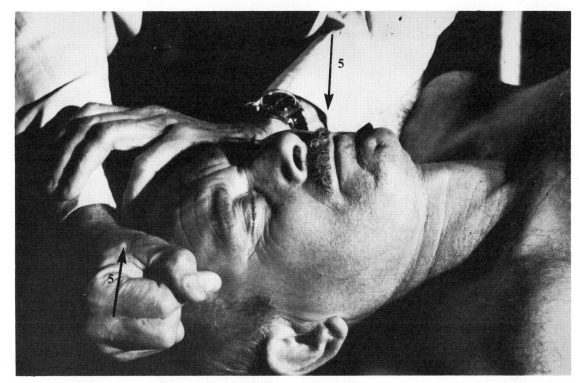

Fig. 9. **Treatment for sphenoid disorders.** The tender point is located over the greater wing of the sphenoid bone on the temple. Note monitoring index finger of the right hand. The temple is tender here and appears more prominent, partly from temporalis tension, but also from eccentric location of the sphenoid bone. This is restored to balance by pushing the opposite sphenoid wing toward the right side. Though the right hand is used to monitor, it is also used for counterpressure over the frontal bone.

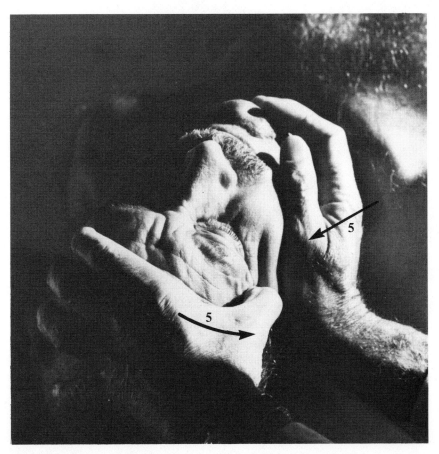

Fig. 10. **Treatment for lateral canthus dysfunction.** The tender point is found in the temporal fossa about 2 cm. posterior to the lateral canthus of the eye. It is treated by upward pressure from the palm of the hand on the lower surface of the zygomatic bone and the zygomatic process of the maxilla. Counterforce is applied by pulling on the top of the frontal bone toward the zygoma.

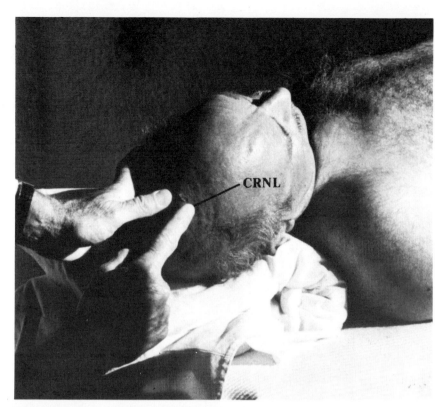

Fig. 11. **Treatment for coronal disorders.** The coronal tender point (CRNL) is located 1 cm. from the anterior medial corner of the right parietal bone (and on it), just behind the coronal suture and just lateral to the sagittal suture. Relief is obtained simply by pressure at the same point on the opposite parietal bone.

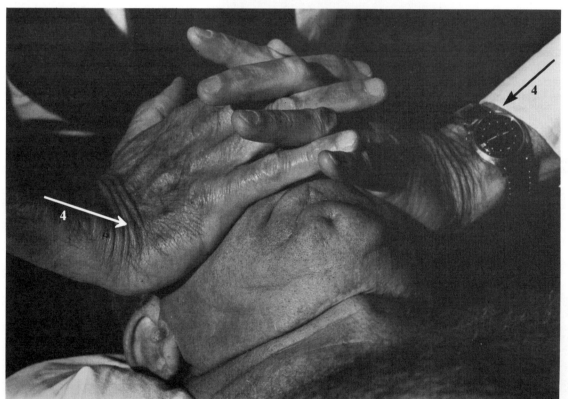

Fig. 12. **Treatment for infraorbital dysfunction.** Tender points are near the site of emergence of the infraorbital nerves. A common, important problem, infraorbital dysfunction is related to many "sinus headaches." Both hands are used in treatment, with pressure applied obliquely medialward and backward from the middle of the palms, which are stretched over the cheek bones. The patient is aware of mild discomfort under the doctor's palms but also of a feeling that the pressure in back of the nose has been relieved. Prompt nasal decongestion often results. Be sure to leave space between the fingers and the patient's nose, so that the patient can breathe.

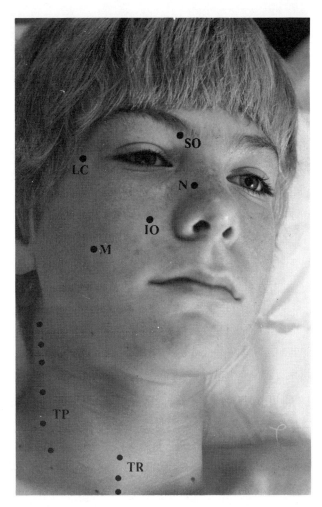

Fig. 13. **Locations of several facial, throat, and neck tender points.** Supraorbital (SO), nasal (N), infraorbital (IO), lateral canthus (LC), and masseter (M) points all are related to cranial suture dysfunctions. The row of tender points found on the anterior surfaces of the tips of the transverse processes (TP) of the cervical vertebrae indicate forward-bending dysfunctions of the cervical intervertebral joints. The row of tracheal tender points (TR) near the center of the anterior neck frequently are associated with unexplained chronic, unproductive cough.

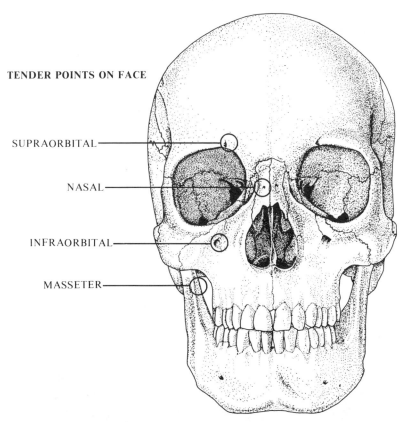

TENDER POINTS ON FACE

- SUPRAORBITAL
- NASAL
- INFRAORBITAL
- MASSETER

47

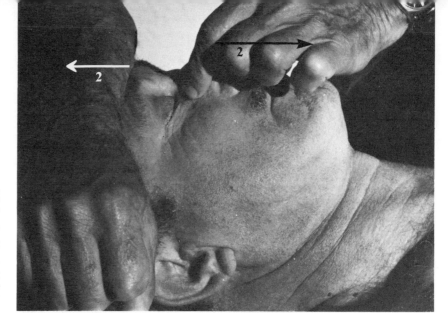

Fig. 14. **Treatment for supraorbital dysfunction.** The tender point is near the site of emergence of the supraorbital nerve. The physician's right forearm is pressed over the frontal bone, pulling it cephalad. The left fingers pinch just above the bridge of the nose and pull caudad.

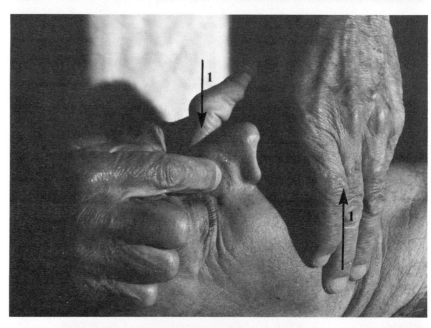

Fig. 15. **Treatment for nasal dysfunction.** The tender point is on the side of the bridge of the nose. Relief is obtained simply by pressing on the opposite side.

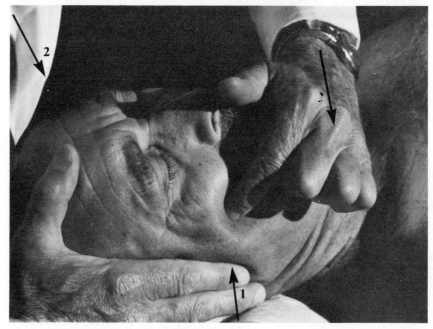

Fig. 16. **Treatment for masseter disorders.** This method is used against two different problems; it relieves most disabilities of the temporomandibular joint and it is likely to relieve dental or mandibular neuritis. The tender point is monitored by the index finger of the left hand, which also pushes on the left side of the point of the chin toward the right. Note that the left top of the head is braced against the doctor's chest. The patient allows his jaw to open 1 cm. Counterforce from the fingers of the right hand is applied, pressing left. When used for temporomandibular disability, the diagnosis can be made by asking the patient to slowly open his jaw. It will be seen to deviate toward the sore side. Again, exaggerate the deformity when treating this dysfunction.

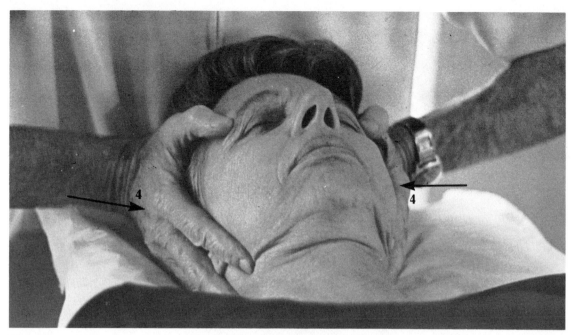

Fig. 17. **Bilateral compression of the head.** This treatment is used at times for bilateral squamosal or bilateral posterior auricular tender points. Sometimes this treatment is given just because it feels good to the patient. The pressure is applied just behind the ears bilaterally.

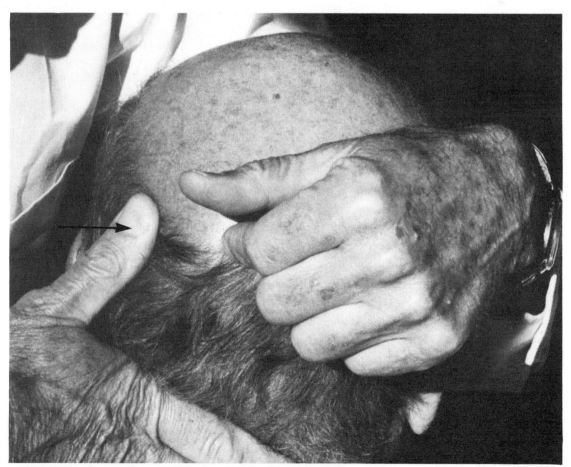

Fig. 18. **Treatment for lambdoidal dysfunction.** This procedure seems too simple (as does treatment of coronal dysfunctions). The tender point is just medial to the lambdoid suture and obliquely up from the inion and out. To treat this dysfunction, just press at the same point on the opposite side. See Figures 4 and 11.

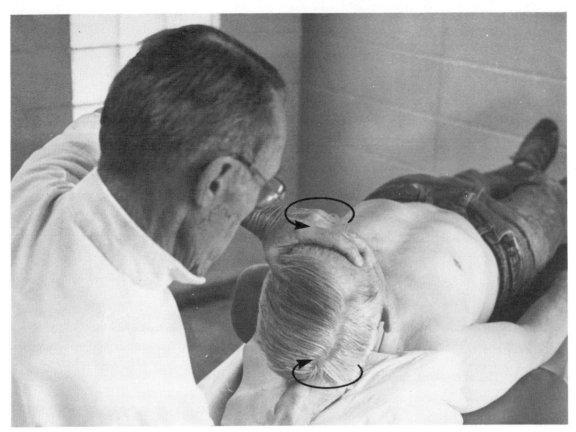

Fig. 19. **Treatment for sphenobasilar torsion.** The tender point is 2 cm. medial to the lambdoid suture above the level of the inion. Treatment is accomplished by counterclockwise rotation applied to the frontal bone (as viewed from the front), with counterrotary force applied to the occipital bone. This causes torsion through an anteroposterior axis.

Fig. 20. **Treatment for zygomatic dysfunction.** The tender point is found just above the zygomatic arch of the temporal bone. It is about 3 cm. anterior to the external auditory meatus. Force is applied from below the arch of the temporal bone. It is directed upward with counterforce pulling the top of the head toward the temporal bone. (This is similar to treatment for the lateral canthus dysfunction, but is applied 4 cm. further posteriorly.)

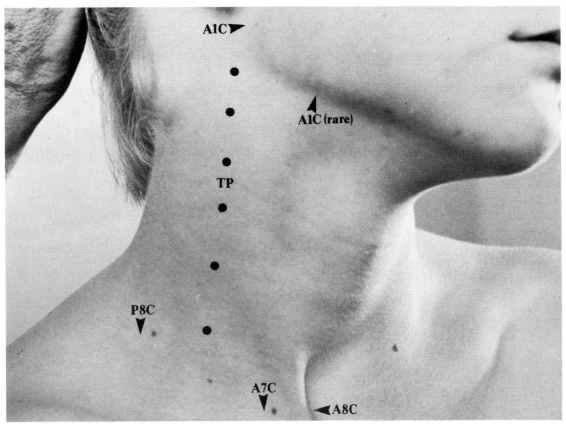

Fig. 21. **Locations of tender points for anterior neck disorders.** The two marked A1C are for the anterior first cervical joint in two different types of dysfunction. The upper tender point is located on the posterior surface of the ascending ramus of the mandible just under the ear; the lower one is on the inner table of the mandible 2 cm. anterior to the angle. In the row marked TP are transverse process tender points. Two eighth cervical level tender points indicate flexion and extension disorders of that joint. The posterior eighth cervical tender point (P8C) is found by deep palpation anterior to the trapezius muscle on the posterior surface of the tip of the transverse process of the seventh cervical vertebra. The anterior eighth cervical tender point (A8C) is located on the medial end of the clavicle. The anterior seventh cervical tender point (A7C) is located 3 cm. lateral to the medial end of the clavicle on its posterosuperior surface.

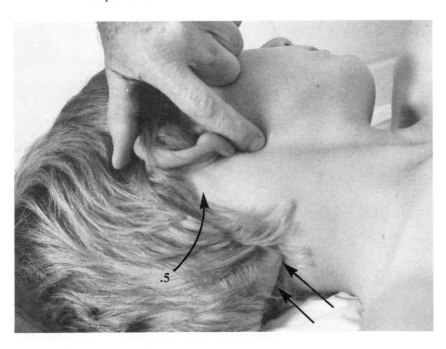

Fig. 22. **Treatment and monitoring of the anterior first and second cervical joint dysfunctions.** These are the most common anterior cervical joint dysfunctions. The tender point for the regular anterior first cervical is the one located on the posterior edge of the ascending ramus of the mandible. The occiput is supported by the right hand, and the head is rotated left with a light force of .5 kg. Note the fingers of the right hand on the occiput. There is slight sidebending left.

Fig. 23. **Treatment used with minor variations for the third cervical dysfunctions and one of the anterior first cervical dysfunctions.** This first cervical tender point is the one located on the inner table of the mandible anterior to the angle. They are both treated with marked forward bending of the upper neck, sidebending (usually toward the right), and rotation left. A third tender point for the first cervical joint (1C, inion) is not shown here; it is found on the medial surface of the main posterior cervical muscle mass attachment to the occiput, 3 cm. below the inion. This tender point is often turned off similarly. (See Figures 4 and 29.)

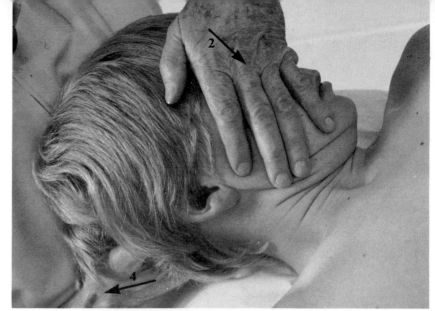

Fig. 24. **Treatment for anterior fourth cervical joint malfunction.** Observe that the head is suspended over the table edge and allowed to extend slightly. The tender point is on the anterior surface of the tip of the transverse process of the fourth cervical vertebra. Action and force of 2 kg. left rotation and left sidebending are used. Although this stretch uses little or no forward bending, it contrasts with the extreme backward bending used for posterior fourth cervical vertebra dysfunction. (See Figure 31.)

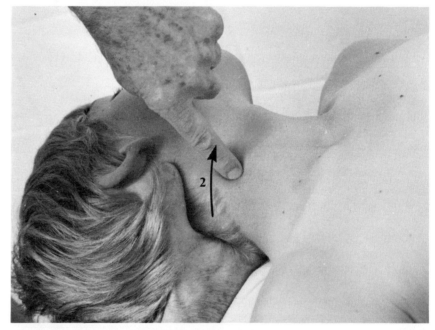

Fig. 25. **Treatment for anterior fifth and sixth cervical joint dysfunctions.** The neck is slightly flexed. Although rotation is the obvious action, probably more of the force used is in sidebending left. The tender points are located on the anterior surface of the tips of the transverse processes of the fifth and sixth cervical vertebrae.

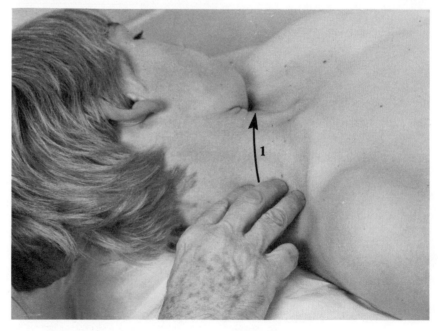

Fig. 26. Treatment for anterior seventh cervical joint dysfunction. The method includes marked flexion of the low neck, which is achieved by supporting the mid-neck with the left hand rather than by pressing on the head. By this means the force is located at the proper joint level. Rotation is to the left and slight; sidebending is to the right and often fairly marked. The tender point is located on the posterosuperior surface of the clavicle, 3 cm. lateral to the medial end.

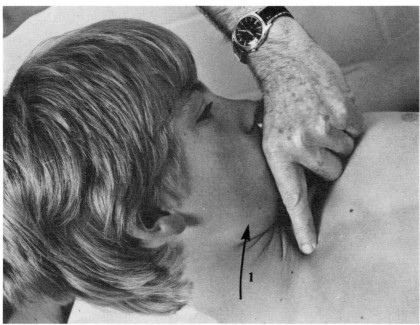

Fig. 27. Treatment for anterior eighth cervical joint dysfunction. The method is slight forward bending with marked left rotation through the eighth cervical joint. Again, even though the obvious action is mostly rotation, the force applied is mostly sidebending left. The tender point is located at the medial end of the clavicle.

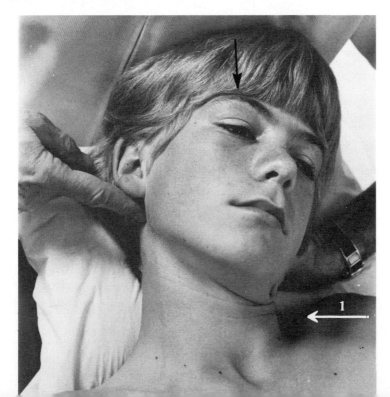

Fig. 28. Treatment for lateral first cervical joint dysfunction. Although this disorder does present a tender point on the tip of either transverse process of the first cervical vertebra, it is one of the dysfunctions easily diagnosed by bony relationships. The transverse process of the first cervical vertebra is approximated more closely to the mastoid process of the temporal bone on one side than the other. The direction is easily palpable. There also is an apparent "sideslip" toward the lateral concavity. Whatever positional deformity is found should be exaggerated in treatment. The action here is simple sidebending, which is performed by applying pressure on one side of the occiput and jaw and counterpressure on the top of the other side of the head. Note that the pressure by operator's chest helps to produce needed sidebending. Evidence of release is mostly from a restored neutral position of the vertebra. The relationship between the side of the bend and the location of the tender point still is not definite to the author.

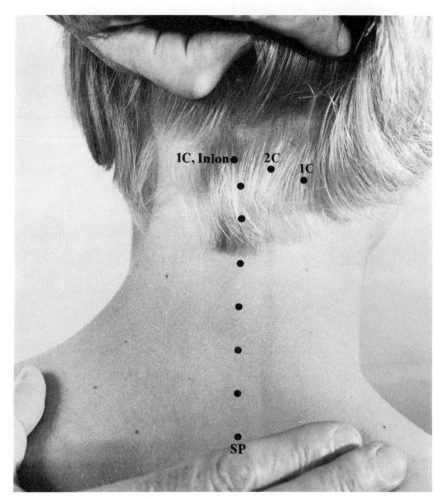

Fig. 29. **Location of tender points for posterior cervical joint dysfunctions.** The inion tender point (1C, inion) is found on the medial side of the large posterocervical muscle mass where it attaches to the occiput approximately 3 cm. below the inion. It is one of three tender points of the forward-bending first cervical joint dysfunctions (see Figure 23). The tender point for posterior second cervical joint dysfunction (2C) is located on the lateral side of the same muscle mass and upper surface of the spinous process of the second vertebra. That area marked 1C is the tender point on the occiput in an area of thin muscle halfway between the muscle mass and the mastoid process. The side of the prominent spinous process of the second cervical vertebra is the site of a tender point for the third cervical joint dysfunction. It is also another posterior tender point for a forward-bending disorder. (See Figure 23.) The column of dots marked SP indicates the spinous process tips.

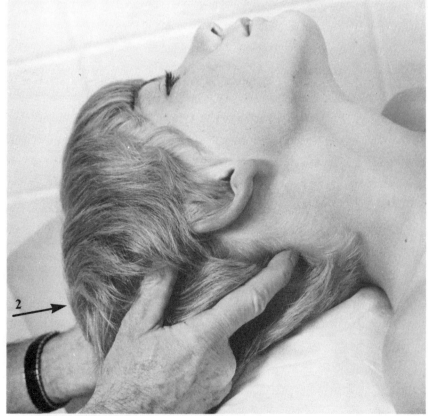

Fig. 30. **Treatment for posterior first and second cervical joint dysfunctions.** Force is used here to localize the stretch mostly at the level of either the first or second cervical joints; it is applied caudadly on top of the occiput. The action is marked extension, slight rotation, and side-bending left. Although the tender point for the third cervical joint is located on the spinous process of the second cervical vertebra, it is irregular and is nearly always relieved in marked forward bending. (See Figure 23.)

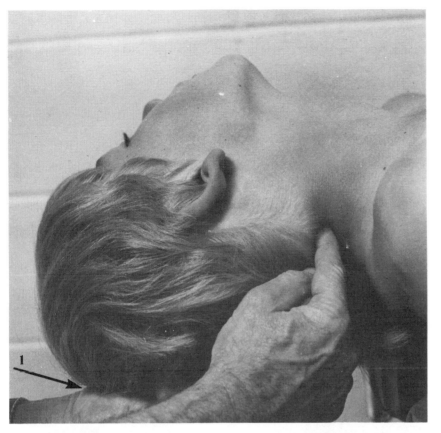

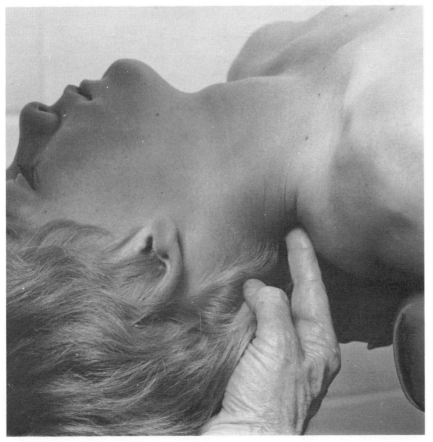

Fig. 31. **Treatment for posterior fourth cervical joint dysfunction.** Once the tender point for this joint is found on the spinous process of the third cervical vertebra, the practitioner must move his monitoring finger laterally to another tender point on the muscle mass. Although this point is less sensitive for discovering the disorder, it is the only one that will remain palpable in the position needed for release. Notice that the patient's head is suspended over the end of the treating table. The localization of force into the fourth cervical joint with posterior dysfunction is aided by force applied caudadly on the occiput (much less than that used for the first cervical joint). This is followed by slight sidebending away and rotation away.

Fig. 32. **Treatment for posterior lower posterocervical and upper thoracic joint dysfunctions.** These are treated similarly to the posterior fourth cervical joint, with minor variation. The head is allowed to hang increasingly further backward in order to locate the force applied into the lower joints for stretching. The posterior first, second, and fourth cervical problems require progressively further backward positioning of the head in relation to the chest and progressive lessening of the caudad force above the occiput. These lower cervical and upper thoracic dysfunctions are treated effectively with just the use of the weight of the head and neck. They also can be treated with the patient in a prone position with the neck extended, but the supine position is more effective, because there is less active "assistance" by the patient. Again sidebend away slightly and rotate away.

Fig. 33. **Tender spots for anterior (forward-bending) dysfunctions in the thoracic area.** The row over the sternum labeled AT indicates anterior thoracic intervertebral joint tender points from the first thoracic tender point down to the ninth thoracic tender point just above and lateral to the umbilicus. The rib interspace tender points (INT) indicate less common rib problems; these will be relieved by the same stretches as the midthoracic intervertebral joint treatments (anterior seventh and eighth thoracic levels). Interspace tender areas (INT) are found on or between the costal cartilage just lateral to the sternum. Note that the direction of the probing force for the anterior first thoracic tender point (A1T) is downward into the center of the suprasternal notch. There are two anterior seventh thoracic tender points; the one indicated by the arrow is located under the costochondral margin. Depressed rib tender points that are released by further depression (toward extreme exhalation) start on the first rib just under the clavicle and close to the sternum and on the second rib in the midclavicular line; those for the third to sixth depressed ribs are found on the anterior axillary line inferior rib margins. The location of the tender point for disorders of the anterior acromioclavicular joint (AAC) is 1 to 2 cm. medial to the tip of the clavicle on its anterior surface.

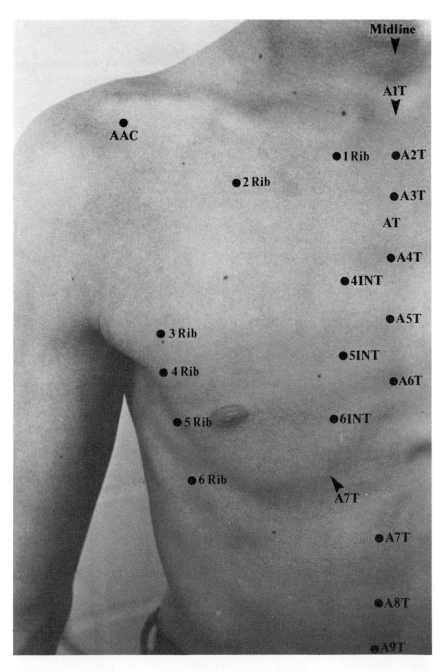

56

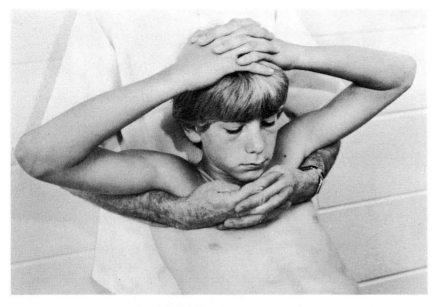

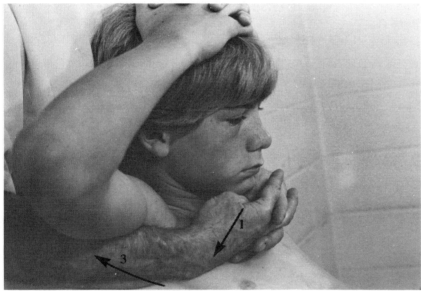

Figs. 34 and 35. **Treatment for the anterior first and second thoracic intervertebral joints in disorders in forward bending.** Raising the arms fixes the force applied into the upper thoracic joints. The physician stands behind the patient, who is seated crossways on the table. The patient leans back against the chest of the doctor and slumps forward. The pressure from the doctor's chest assists in forced flexion.

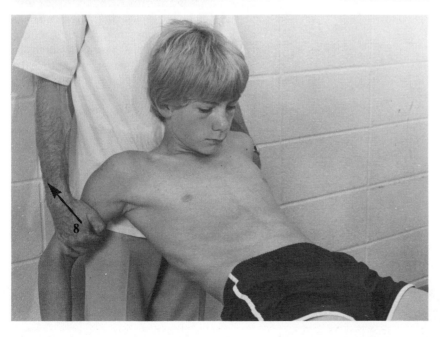

Fig. 36. **Treatment for anterior third and fourth thoracic joint dysfunctions.** Use the same general position of the patient as for anterior first and second thoracic tender points. Here, however, the arms are pulled backward. The force needed in this area is greater than that applied for first and second joint disorders. The pressure from the physician's abdomen assists the forced flexion.

57

Fig. 37. (right) **Treatment for the anterior fifth and sixth thoracic joint dysfunctions.** This difficult treatment requires considerable force applied by the doctor's hands over the sternum. The force is applied so as to cause as much midthoracic flexion as possible without straining the more flexible thoracolumbar area. This is aided by a little pulling force cephalad. This tender point is very common and should be suspected in patients who awake tired in the morning. Note that action is enhanced by pressure from the doctor's chest on the upper thoracic spine.

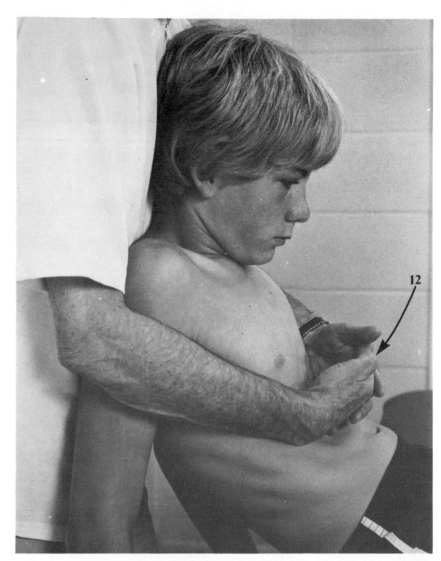

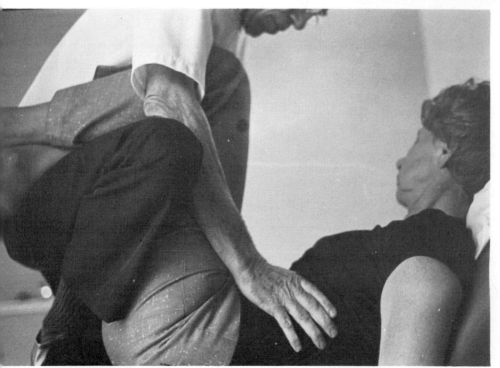

Fig. 38. (left) **An alternate method of dealing with the common and difficult anterior fifth, sixth, seventh, and eighth thoracic intervertebral joint dysfunctions.** Although this is usually effective, it may be necessary also to do the treatments suggested in captions for Figures 37, 39, and 40. The head of the table is raised (see Figure 51) and the patient's body is placed so that there is room for flexion at the desired spinal level. The midthoracic treatments require, in addition to the marked flexion achieved by use of the legs and thighs, an additional pressure of 2 to 4 kg. applied by the hands of the physician over the lower chest.

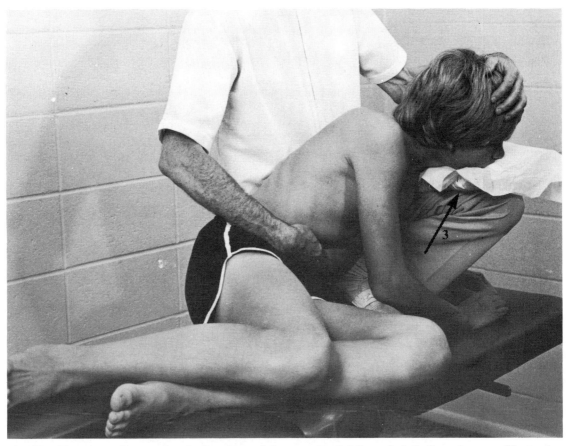

Fig. 39. **Treatment for anterior dysfunctions of the seventh, eighth, and ninth thoracic areas.** The tender points for these segments are found near the midline from a point just below the xiphoid on down to the umbilicus. The patient is seated on the table leaning left, with his left axilla supported by the thigh of the operator, who stands behind him. The patient's body is rotated left, but sidebent right with little forward or backward influence. The tender point is monitored by the operator's free hand. The interspaces indicated on Figure 33 are treated in the same way. These interspace tender points indicate another type of rib dysfunction. They are more commonly found just lateral to the sternum at their corresponding levels — the fifth and sixth interspaces.

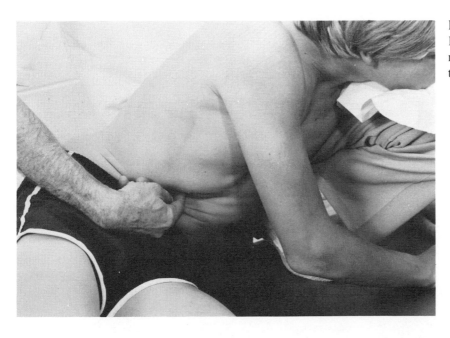

Fig. 40. **A closer view of Figure 39.** It shows the monitoring of the anterior seventh thoracic tender point in the epigastrium.

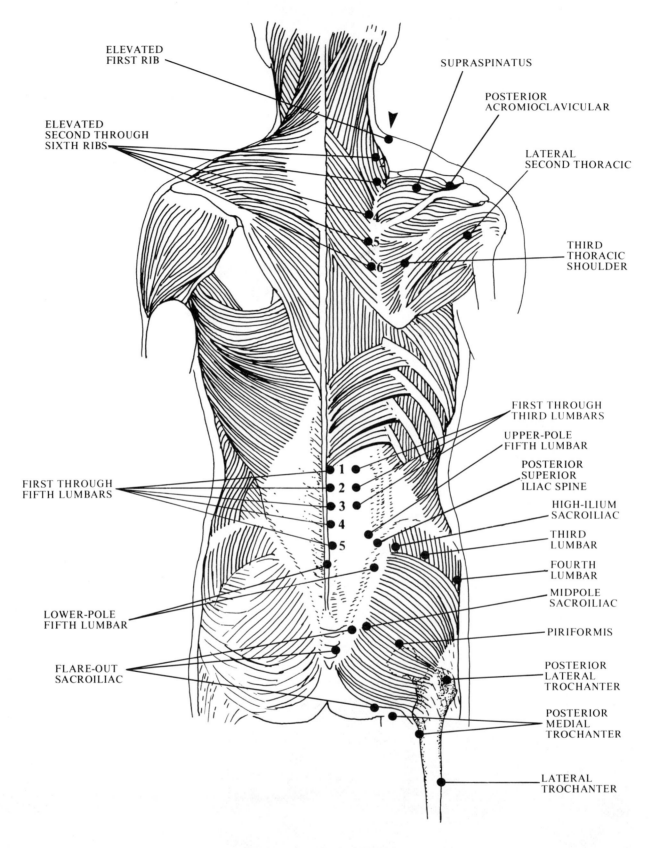

POSTERIOR TENDER POINTS

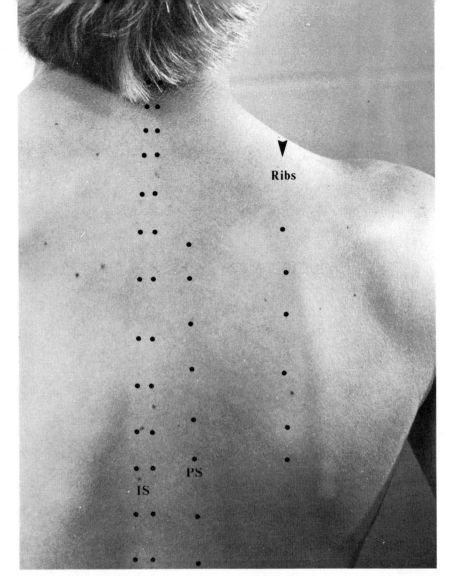

Fig. 41. (left) **Tender points of the posterior thorax.** These are locations (interspinal [IS], paraspinal [PS], and the rib angles) for tender points associated with extension dysfunctions of intervertebral joints, sidebending dysfunctions, and ribs that are more comfortable if elevated.

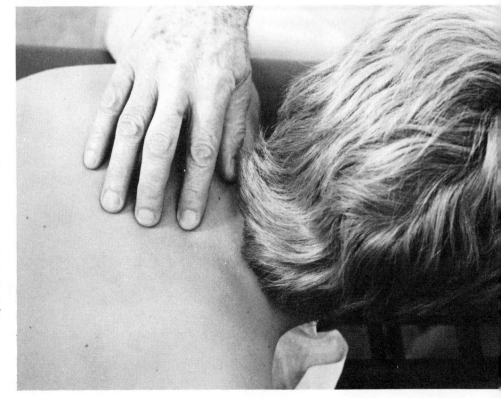

Fig. 42. (right) **Prone position treatment of lower posterocervical and upper posterior first and second thoracic joints.** (See Figure 32 for treatment in the supine position.) The head is supported by the doctor's left hand holding the chin. The operator's left forearm is held along the right side of the patient's head for better support. (The right hand monitors tender points on the right side of the spinous processes.) The forces applied are mostly extension, with slight sidebending left and rotation left. Although the fingers of the monitoring hand in the picture appear to be on the left side of the body, they are actually just to the right of the spines of the vertebrae. The patient's arms are suspended over the side of the table.

Fig. 43. **Similar treatment and monitoring of tender points for the posterior third, fourth, and fifth thoracic joints.** The principal difference is the placement of the patient's arms, which are now extended above his head. This helps to localize forces to the involved area. Tender points in this area are usually found close to the spinous processes.

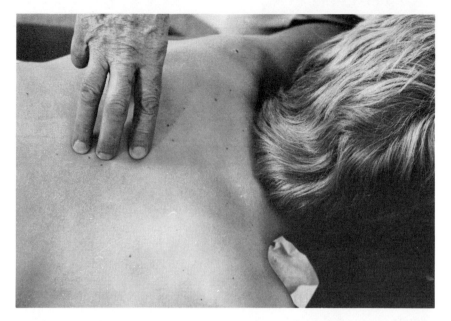

Fig. 44. **Treatment for many posterior joint dysfunctions of the sixth thoracic through the second lumbar joint.** The operator grasps the patient's arm near the axilla without undue pulling of sensitive skin. The direction of force is mostly sidebending left, with mild rotation right. When used for midthoracic problems, the effect can be improved by placing the head left and rotating it right. The right hand in this figure does not show tender point monitoring, but the area treatable with this method.

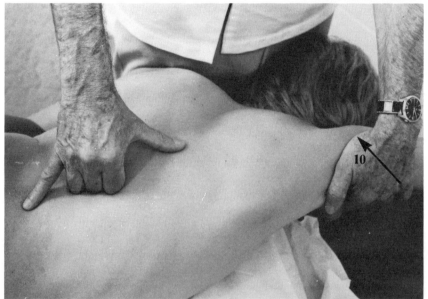

Fig. 45. **Treatment for posterior mid- and lower thoracic dysfunctions with tender points in the midline on or near the spinous processes.** Although many of the mid- to lower thoracic joint dysfunctions have tender points more laterally and are treated with much sidebending, these problems are treated with extension (most easily performed by elevating the cephalic end of a McManis table) and rotation by pulling back on the pelvis. The spinous processes at the site of the tender point indicate rotation directions. Treatment may be accomplished on a flat table by placing a cushion or large pillow under the shoulders and chest of the prone patient. Rotation is easily done, exaggerating the abnormal rotation found by palpating the spinous processes. The operator reaches across to grasp the pelvis at the anterosuperior spine and gently rotates it.

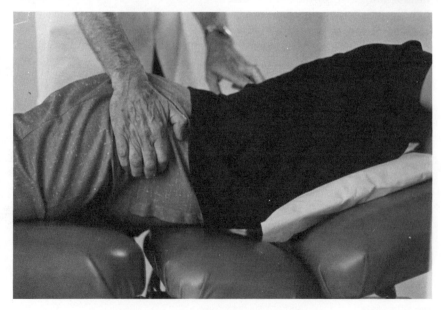

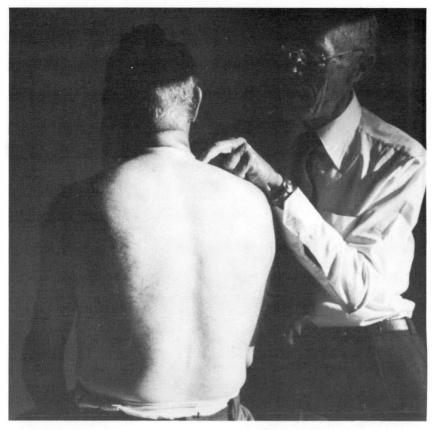

Fig. 46. **Treatment for elevated first rib.** This photo shows the location of the tender point beneath the margin of the trapezius at the side of the neck. The position of treatment has the head and neck slightly extended, moderately rotated right and slightly sidebent left. This maneuver is surprisingly effective although it seems to be doing very little. Although this photo shows the doctor before the patient, this is only for the use of this illustration. In fact, the physician stands behind the seated patient, whose feet are hanging off the table. The physician places his left foot on the table and drapes the patient's left arm over the physician's left thigh. The patient is encouraged to lean a little weight on it.

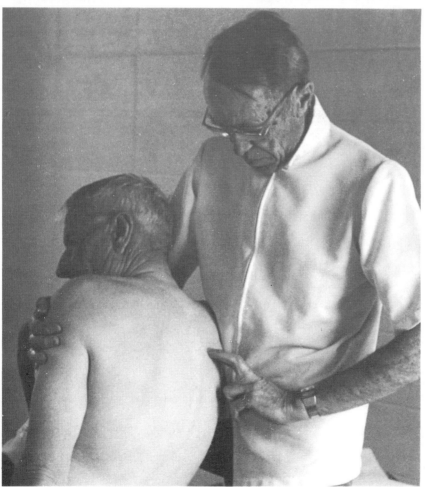

Fig. 47. **Treatment for elevated ribs.** The operator stands behind the patient, who sits on the table, leaning right. His right axilla is supported by the thigh of the doctor. The patient has one or both feet on the table to his left, beside his hips. By this method the doctor can produce fairly marked left sidebending and rotation at the upper thoracic level. Although this maneuver is used for all elevated rib dysfunctions, the treatment for the second rib requires rotation of the neck and head as well. For the other segments this appears to be unimportant.

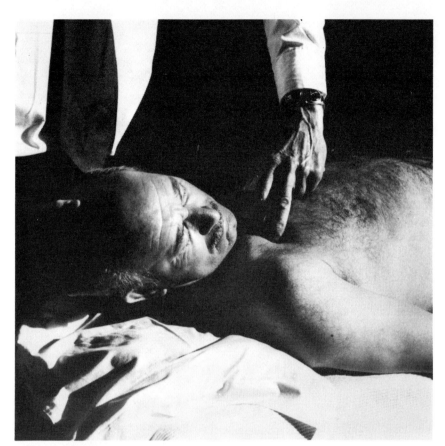

Fig. 48. (right) **Treatment for depressed first and second ribs.** The patient is supine and the operator sits or stands at the head of the table. His fingers monitor the tender points of both ribs. The action is sidebending right, rotation right, and mild flexion.

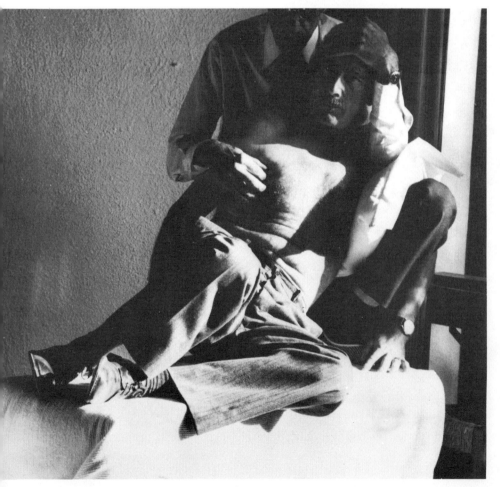

Fig. 49. **Treatment for depressed third to sixth ribs.** The tender points are found in the anterior axillary line. With the patient seated, the physician stands with his left foot on the table. The patient's left arm is thrown over a pillow over the doctor's left thigh. Notice that although the patient has his feet on the table to his right, he leans to his left. The force brought to bear in the upper thoracic area is sidebending right. There is a considerable amount of right rotation and slight forward bending. (This identical position will relieve left elevated rib disorders.)

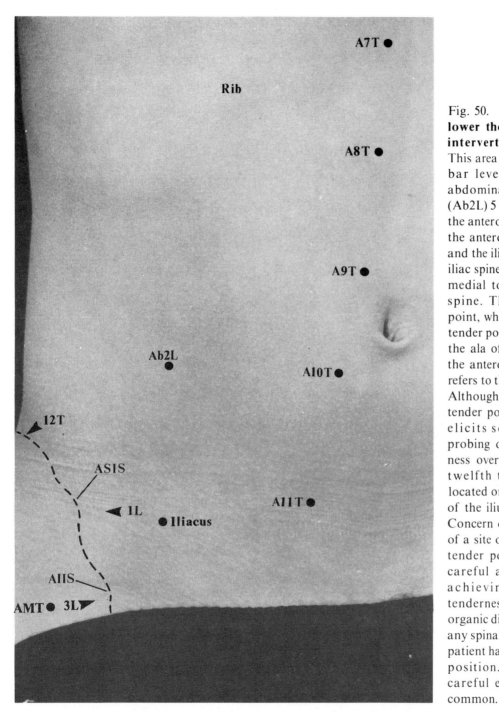

Fig. 50. **Tender points for anterior lower thoracic and upper lumbar intervertebral joint dysfunctions.** This area goes through the first lumbar level (1L) and includes the abdominal second lumbar vertebra (Ab2L) 5 cm. lateral to the umbilicus, the anterosuperior iliac spine (ASIS), the anteroinferior iliac spine (AIIS), and the iliac crest. The anteroinferior iliac spine is located 4 cm. below and medial to the anterosuperior iliac spine. The iliacus muscle tender point, which is probably a hip socket tender point, is located in the fossa of the ala of the ilium 7 cm. medial to the anterosuperior iliac spine. AMT refers to the anteromedial trochanter. Although palpation of the anterior tender points in the abdominal wall elicits some information, deep probing often reveals sharp tenderness over the vertebral bodies. The twelfth thoracic tender point is located on the inner table of the crest of the ilium in the midaxillary line. Concern over the possible mistaking of a site of true organic disease for a tender point can be dealt with by careful assessment of success in achieving substantial relief of tenderness. The tenderness of true organic disease will not be relieved by any spinal stretch, especially after the patient has been returned to a neutral position. Doubtful cases require careful evaluation. They are not common.

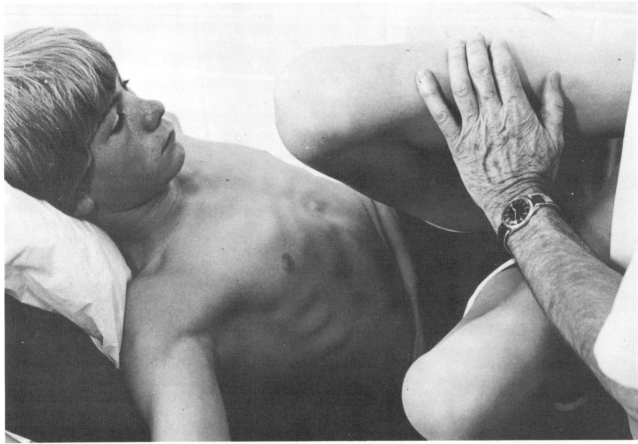

Fig. 51. **Treatment for forward-bending joint dysfunctions from the ninth thoracic through the first lumbar levels.** This one procedure is usually effective for any of this group. To permit the supine spine to flex at the thoracolumbar region, a table capable of being raised at one end is desirable. A flat table may be used if a large pillow is placed under the patient's hips, raising them enough to permit flexion to reach the desired level of the spine. With the patient supine, the physician raises the patient's knees and places his own thigh below those of the patient. By applying cephalad pressure on the patient's thighs, he produces marked flexion of the patient's thoracolumbar spine. Usually, the best results come from rotation of the knees moderately toward the side of tenderness.

These joint dysfunctions account for many low-back pains that are not associated with local tenderness over the vertebra posteriorly. The pain is referred from the anterior dysfunction into the low-lumbar, sacral, and gluteal areas. Treatment directed to the posterior pain sites of these dysfunctions rather than to the origins of the pain has been disappointing.

The position is marked flexion through the joint involved, with mild rotation away and slight sidebending toward (classic description of rotation). Because the physician moves the lower part of the patient's body under the upper part, he moves the knees to the right.

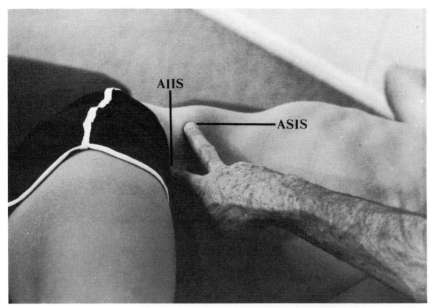

Fig. 52. **Treatment for regular anterior second lumbar dysfunction, a common problem.** The tender point is found on the inferior medial surface of the anteroinferior iliac spine (AIIS). The pelvis is rotated 60 degrees to the left. This causes right rotation of the second lumbar vertebra. Rotation is just the reverse of that shown in Figure 54. Both disorders are usually helped by sidebending left. Both treatments need to avoid pelvic strain from leverage on thighs. A pull on the upper leg (the right for anterior second lumbar dysfunction) is enough. The anterosuperior iliac spine (ASIS) is shown only as a landmark.

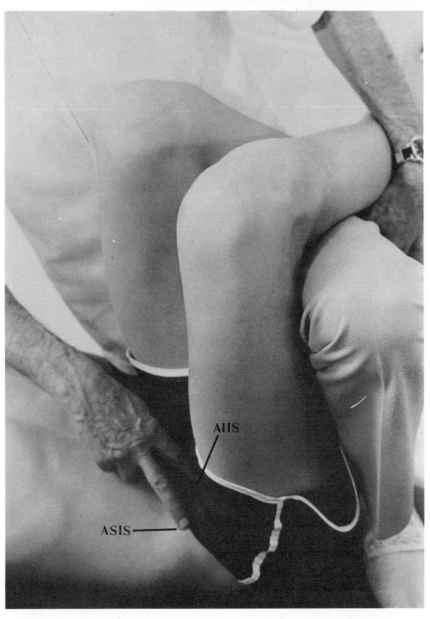

Fig. 53. **Treatment for anterior third and fourth lumbar intervertebral joint dysfunction.** The anterior third lumbar tender point is on the lateral surface and the anterior fourth lumbar tender point is on the inferior surface of the anteroinferior iliac spine (AIIS). They are treated identically. The landmark anterosuperior iliac spine (ASIS) is again pointed out. Note that the physician has a foot on the table to support the patient's legs. Although the patient's hips are flexed, the force applied is sidebending through the lumbar vertebrae to the left.

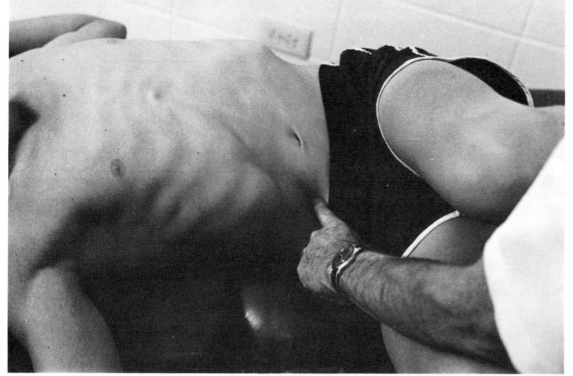

Fig. 54. **Treatment for abdominal second lumbar dysfunction.** The tender point is relieved by flexing the hips 90 degrees and rotating the pelvis about 60 degrees toward the right. Sidebending left is accomplished by raising the feet toward the ceiling and to the patient's left more than the knees. Although this relieves the dysfunction, the positioning may cause an adduction strain through the flexed hip joint. This can be avoided by a pull behind the knee of the upper leg, which results in traction through the thigh.

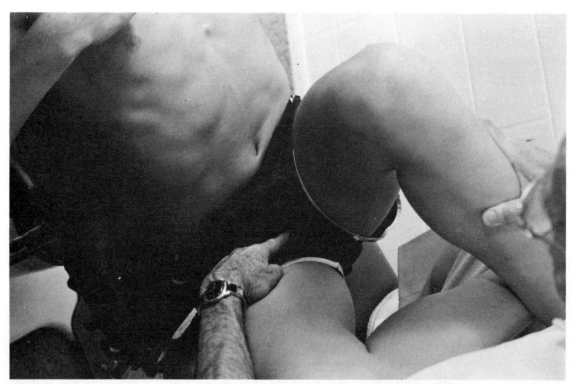

Fig. 55. **Treatment for anterior fifth lumbar joint dysfunction.** The patient lies supine on a flat table, with his thighs flexed to 60 degrees. His feet are crossed over the thigh of the physician. The knees are moved 20 degrees right; this causes the lower spine to be sidebent left. The tender point is on the front of the pubic bone about 1 cm. lateral to the symphysis pubis. Occasionally the examining physician will find bilateral lower-pole fifth lumbar tender points. In this case he will also find an anterior fifth lumbar tender point. Release of the anterior point will usually release the lower pole fifth lumbar areas.

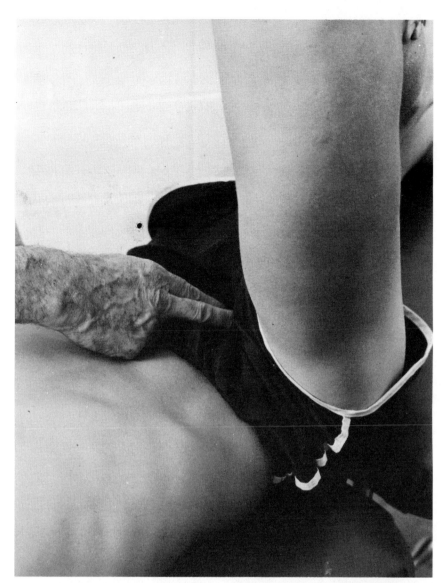

Fig. 56. **Treatment for low-ilium sacroiliac dysfunction.** Often the best means of diagnosing this problem is location of this tender point on the superior border of the lateral ramus of the pubic bone. The thigh is flexed nearly straightforward to about 40 degrees to release the tender point.

Fig. 57. **Variation of the sacroiliac disorder shown in Figure 56, with a lateral flaring out of the top of the ilium in relation to the sacrum.** It is treated similarly, but with marked external rotation of the femur and moderate abduction. The tender point here is located in the perineum on the inferomedial surface of the descending ramus of the pubic bone. Note that the knee is 40 degrees lateral and the foot is at the midline. Flexion is about 110 degrees from straight.

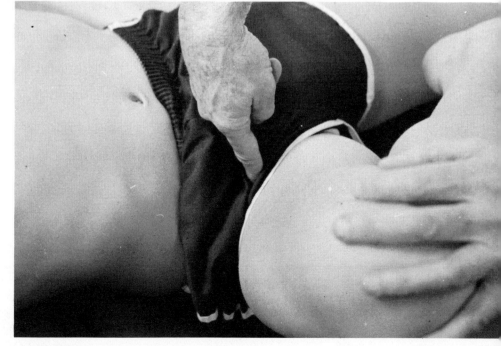

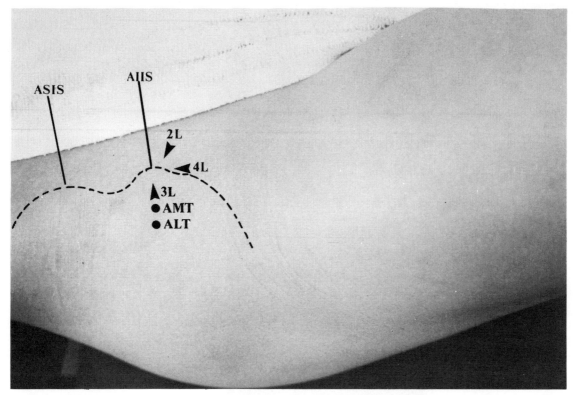

Fig. 58. **Tender points in the groin.** Since this photo does not show the outlines of the pelvis wall, they have been sketched in. In a small area around the anteroinferior iliac spine (AIIS), are four tender points: The very common anterior second lumbar tender point (2L) is medial and inferior to it; the third lumbar tender point (3L) is lateral to it; the fourth lumbar tender point (4L) is beneath it; and, 1 cm. lateral to the third lumbar tender point is the anteromedial trochanter (AMT). This is a hip joint dysfunction. The anterolateral trochanter tender point (ALT) is just lateral to the AMT.

Fig. 59. **Treatment for anterolateral trochanter dysfunction (ALT), another hip joint dysfunction.** The thigh must be flexed 90 degrees to permit deep probing in the site 3 cm. below and slightly lateral to the anterosuperior iliac spine (ASIS). The dysfunction is treated in fairly marked flexion, moderate abduction, and little or no external rotation of the femur.

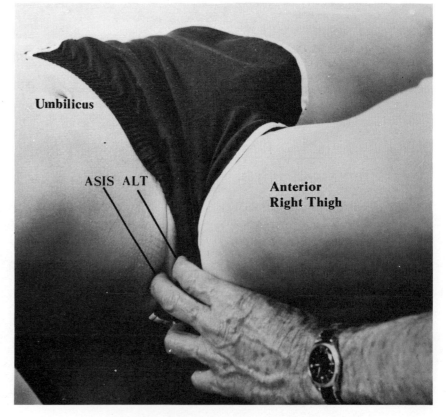

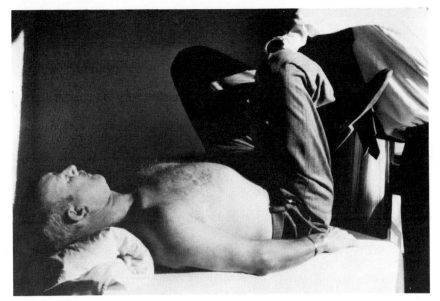

Fig. 60. **Treatment for iliacus dysfunction.** This is probably another hip socket tender point. It is found deep in the lower quadrant over the iliac fossa. It is best treated with the patient supine. Both knees are in marked flexion and there is marked external rotation of the thighs. Note that the knees are separated and the feet are crossed.

Fig. 61. **Treatment for inguinal ligament dysfunction, another hip socket dysfunction.** The inguinal ligament tender point (IL) is located on the lateral border of the pubic bone just caudal to the attachment of the inguinal ligament. It is treated with the patient supine and the doctor standing on the left side. The right knee is flexed 90 degrees and crossed under the left leg. The doctor's left hand is pushing laterally on the patient's right ankle to produce internal rotation of the right femur.

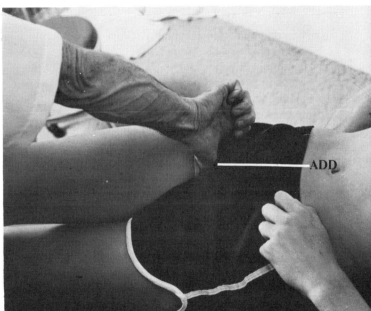

Fig. 62. **Treatment for adductor dysfunction.** The tender point for this hip socket dysfunction is located in the tender adductor muscle (ADD) near its origin on the pubic bone. It is simply released by marked adduction and slight flexion of the hip. The physician stands on the side opposite the side of the dysfunction and draws the leg medially while monitoring for relaxation of the tender point.

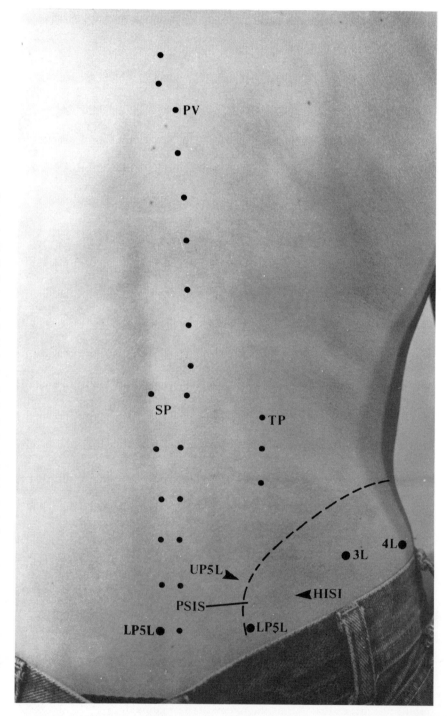

Fig. 63. **Tender points for posterior thoracic and upper lumbar dysfunctions.** Whereas the upper five thoracic areas often have the most sensitive points at or near the spinous processes (SP), the lower thoracic tender points usually are better located paravertebrally or just laterally to the spinous processes. (See line of tender points marked PV.) At the thoracolumbar area there usually are better tender points found over the posterior tips of the transverse processes (TP). Occasionally the spinous process tender points will be the most tender, as they are in the five upper thoracic areas. The postero-superior iliac spine (PSIS) is easily palpated and makes an ideal landmark for several tender points. On its superior medial surface is the common upper-pole fifth lumbar tender point (UP5L). Approximately 2 cm. straight below the UP5L lies a different type of fifth lumbar tender point, the lower-pole fifth lumbar (LP5L). Three cm. lateral to the PSIS, located by probing medially, is the high-ilium sacroiliac tender point (HISI). In this dysfunction the ilium is too high and flared in superiorly in relation to the sacrum. The tender point for the fourth lumbar dysfunction (4L) is found just posterior to the tensor muscle of fascia lata about 4 cm. below the crest of the ilium. About halfway between the PSIS and the fourth lumbar tender point is the third lumbar tender point (3L). An alternate lower-pole fifth lumbar tender point (LP5L), often the better tender point for this dysfunction, is on the central promontory of the sacrum in the midline. (There are often spinal and paravertebral tender points for these problems, but they are usually deep and less discrete and are difficult to monitor accurately.)

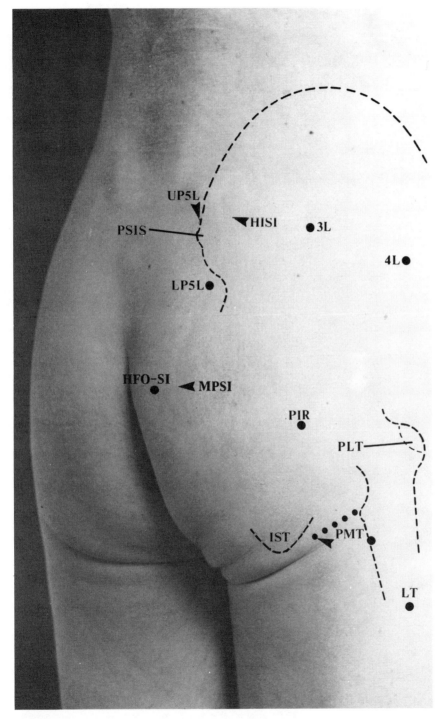

Fig. 64. **Tender points found on the buttocks.** The ilium, the ischial tuberosity, and the greater trochanter of the femur have been sketched in as landmarks, and, starting on the upper left, the posterosuperior iliac spine (PSIS) is the principal landmark. Three cm. lateral to it is the location of the high-ilium sacroiliac tender point (HISI). Two cm. caudad to the PSIS is the tender point for the lower-pole fifth lumbar area (LP5L). More laterally are the tender points for the third and fourth lumbar dysfunctions (3L, 4L). The high flare-out sacroiliac tender point (HFO-SI) is located 10 cm. caudad to the PSIS and slightly medial. The midpole sacroiliac tender point (MPSI) is in an indentation in the buttock musculature and is best found by directing the palpating finger medially. The piriformis muscle tender point (PIR) is about 8 cm. medial and slightly cephalad to the easily palpated greater trochanter. The posteromedial trochanter (PMT) tender point is found in an area from 4 cm. caudad to the trochanter on the posteromedial surface of the shaft of the femur to the posterolateral surface of the ischial tuberosity (IST). The lateral trochanter tender point (LT) is on the lateral surface of the shaft of the femur. Although there is a long row of these on the femoral shaft, the best one is about 13 cm. caudad to the trochanter. A common hip socket dysfunction tender point is located on the superior lateral aspect of the posterior surface of the greater trochanter (PLT).

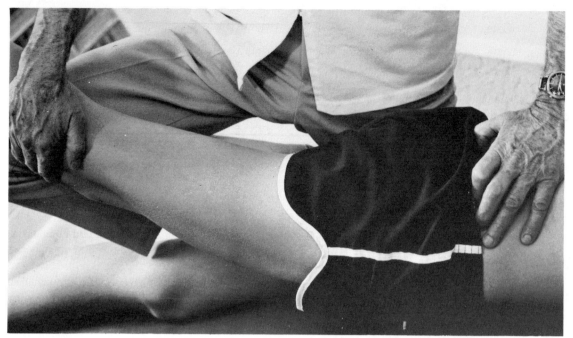

Fig. 65. **Treatment for third, fourth, and upper pole fifth lumbar intervertebral joint dysfunctions.** The lower leg of the patient is supported partially by the knee of the operator. As to modification for treatment of these three levels, the third lumbar area requires more rotation and less extension than the upper pole fifth lumbar area; the fourth lumbar area requires motion in between. All levels usually require some adduction of the femur. Variations in rotation and extension are managed by the placement of the hand lifting the thigh. The higher the hand, the greater the rotation and the less extension.

Fig. 66. **Treatment for lower-pole fifth lumbar dysfunction (LP5L).** The patient lies prone with his right thigh suspended over the side of the table. Positioning of the patient on the table is necessary so as to permit about 20 degrees of rotation of the pelvis. The doctor is seated beside the table and guides the leg into slight adduction. If placement of the patient is correct, the patient's knee will move slightly medially. The doctor's third finger points to the PSIS landmark.

Fig. 67. **Treatment for high-ilium sacroiliac dysfunction.** The tender point (HISI) in the illustration is monitored by the third finger; the PSIS landmark is indicated by the index finger. Note that the doctor's thigh supports the weight of the patient's leg. The action for treating this dysfunction is mainly extension, with slight abduction of the thigh.

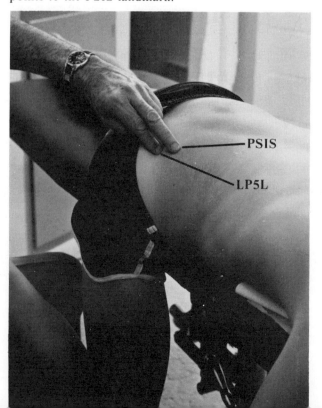

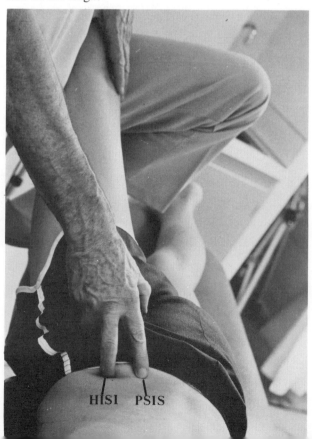

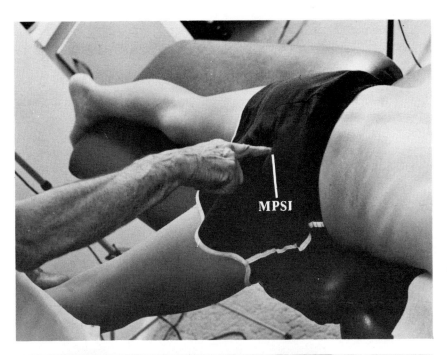

Fig. 68. **Treatment for midpole sacroiliac dysfunction (MPSI).** This is simply abduction of the thigh. Fine tuning consists of slight flexing or extending of the thigh. Note the direction of the probing finger medialward.

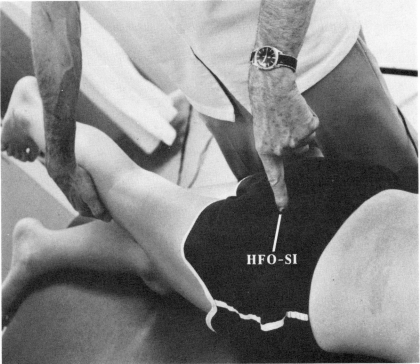

Fig. 69. **Treatment for high flare-out sacroiliac dysfunction (HFO-SI).** The thigh is extended just enough to allow adduction of the leg behind the other thigh. The tender point is monitored 10 cm. below and slightly medial to the posterosuperior iliac spine. This disorder causes the great majority of coccygodynia.

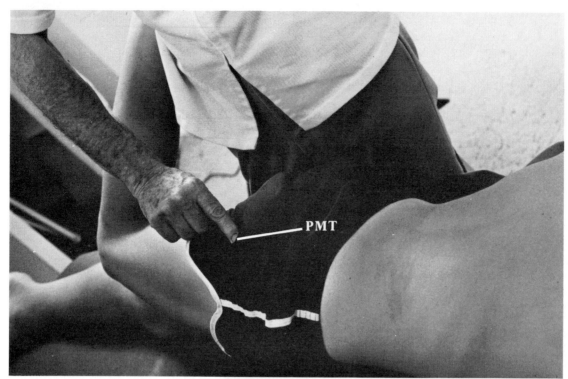

Fig. 70. (above) **Treatment for posteromedial trochanter dysfunction (PMT).** In many cases this hip socket reflex dysfunction is culpable but the sacroiliac joint is blamed mistakenly. The doctor stands on the left side and first pins the patient's right ankle between the doctor's elbow and chest. The physician's left hand reaches around behind himself and grasps the ankle. This frees the doctor's right hand to return to monitoring the changes in the tender point. There are several ways of doing this treatment, but this is the only one that is comfortable to the doctor.

Fig. 71. (right) **Treatment for posterolateral trochanter dysfunction (PLT).** This is another common hip socket reflex dysfunction responsible for much hip pain, with tenderness on the posterosuperior lateral surface of the greater trochanter. Treatment includes extension, marked external rotation, and abduction. Note that the operator's knee is thrust beneath the trochanter in front. Because of a shape like the head of a cane, the femur may be rotated externally by pressure from the front at this point.

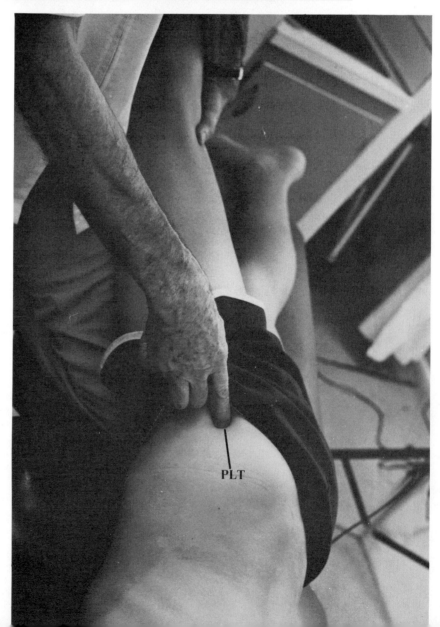

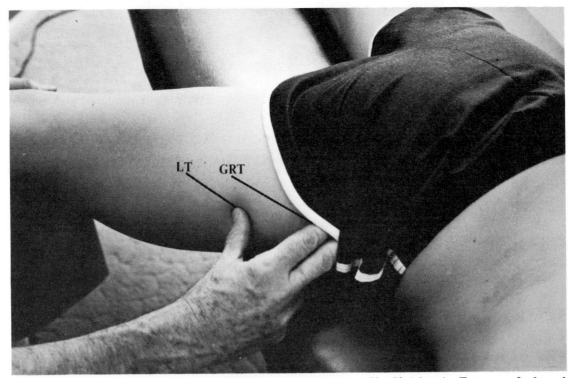

Fig. 72. (above) **Treatment for lateral trochanter dysfunction (LT).** The tender spot is 12 cm. below the greater trochanter (GRT) on the lateral side of the shaft of the femur. This relatively uncommon femoral joint disorder is treated in abduction and usually with a little flexion. The doctor is seated, with the patient's knee resting on his thigh.

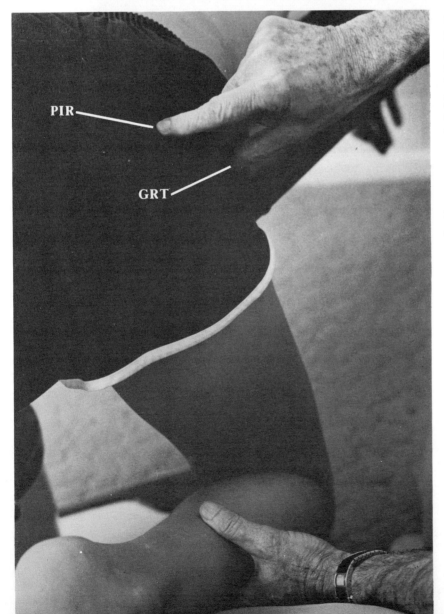

Fig. 73. (left) **Treatment for piriformis muscle dysfunction.** The piriformis tender point (PIR) is located with the aid of the greater trochanter landmark (GRT). Treatment is similar to that for lower-pole fifth lumbar dysfunction, except that flexion of the thigh is increased to 120 degrees from straight. Also, abduction of the femur is used instead of adduction.

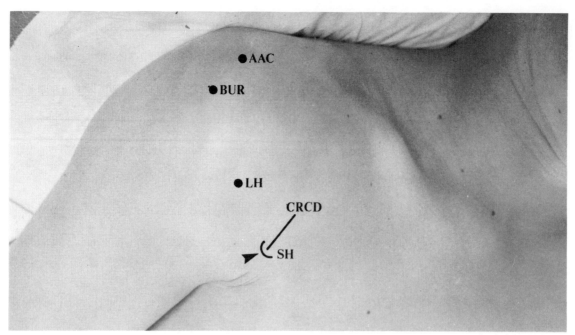

Fig. 74. **Tender points for anterior shoulder joint dysfunctions.** The anterior acromioclavicular tender point (AAC) is on the anterior surface of the distal end of the clavicle. The "infamous bursa" tender point (BUR) is best palpated with the supine patient's arm flexed 90 degrees; this procedure releases enough tension of the superficial muscles so that this none-too-common tender point may be found. The location of this tender point is beneath the acromium process and above and lateral to the long head of the biceps. Although much importance is given to this bursal dysfunction, it is relatively uncommon and is easily relieved with a shoulder stretch. The location of the long head of the biceps muscle tender point is marked LH. The site of the attachment of the short head of the biceps muscle (SH) to the inferolateral surface of the coracoid process (CRCD) of the scapula is also a shoulder tender point. (For description of treatment of SH see Fig. 87.)

Fig. 75. **Treatment for dysfunction of the anterior acromioclavicular joint.** With the patient supine, the doctor stands at the left and draws the patient's right arm obliquely across and downward. The direction varies in different cases, from a straight caudad move to a pull toward the four o'clock position. This view is toward the superior portion of the shoulder. The subject is supine and his neck is at the left border of the picture.

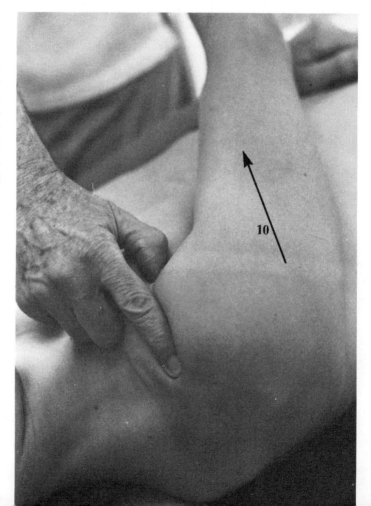

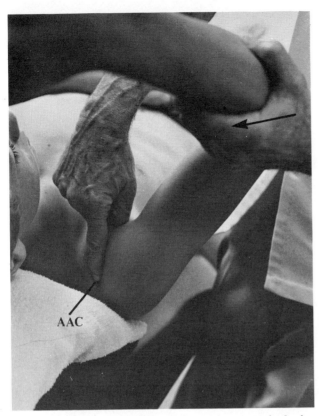

Fig. 76. **Alternate treatment for anterior acromioclavicular joint dysfunction (AAC).** Occasionally this is more effective than the treatment shown in Figure 75. The upper arm of the supine patient is raised to about a vertical position, as indicated by monitoring.

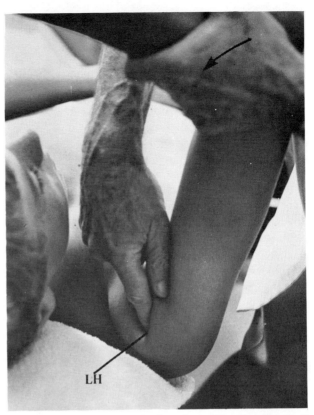

Fig. 77. **Treatment for dysfunction of the long head of the biceps muscle (LH).** This problem is one of many humeral joint dysfunctions. Treatment consists of maximum shortening of the long head of the biceps as indicated by monitoring tissue changes and tenderness.

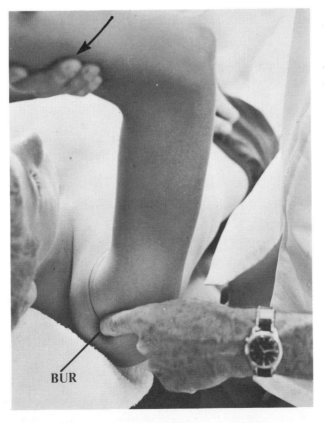

Fig. 78. **Treatment for shoulder bursal dysfunction (BUR), another reflex dysfunction of the humeral joint.** The doctor flexes the shoulder joint of the patient near the position of comfort before this tender point becomes easily palpable. Then, slight modification completes the treatment.

Fig. 79. Two axillary shoulder joint tender points. The pencils probe deeply into the axilla. The upper one reaches to the second rib (2R) in the midaxillary line. This is not a shoulder joint problem per se, but a depressed second rib, which is comfortable only when depressed and which causes apparent shoulder pain when the arm is markedly abducted. It is important as a commonly missed source of shoulder stiffness. The other pencil probes the anterior surface of the scapula and a tender point on the subscapularis muscle (SUB).

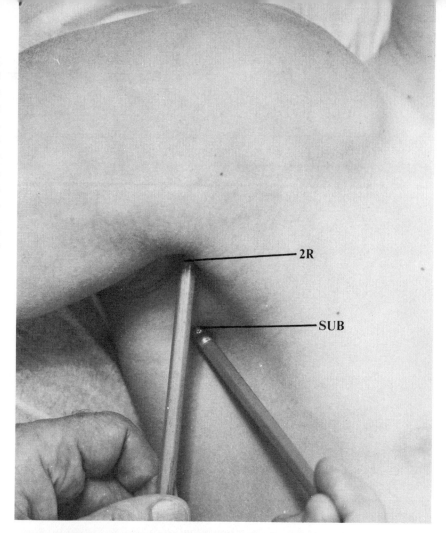

Fig. 80. Treatment of latissimus dorsi dysfunction. The index finger is on the tender point of the latissimus dorsi (LD). It is located more anteriorly on the shaft of the humerus than is the tender point of the adductor shoulder. The middle finger is on the adductor tender point. It is very difficult to probe the undersurface of the insertion of the latissimus dorsi, but one can approach it closely enough to elicit the pain of its tender point. Fingers dip deeply into the axilla and then push forward over the medial surface of the humerus. The tender points for the adduction or "frozen" shoulder are deep in the axillary fossa on its lateral wall. The patient with this disorder cannot move his elbow off his chest. The position shown here also is the treatment for latissimus dorsi dysfunction. Again, relief seems to come from maximum shortening of the muscle containing the tender point. To do this, the patient's arm is pulled caudad and backward to about the seven o'clock position and internally rotated. The force is estimated at 10 kg., most of which is traction. Treatment for subscapularis dysfunction is the same, but without traction.

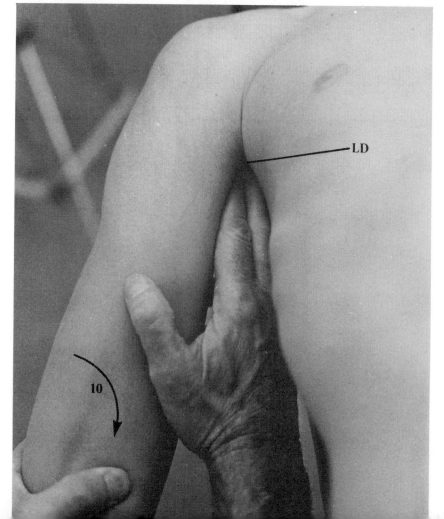

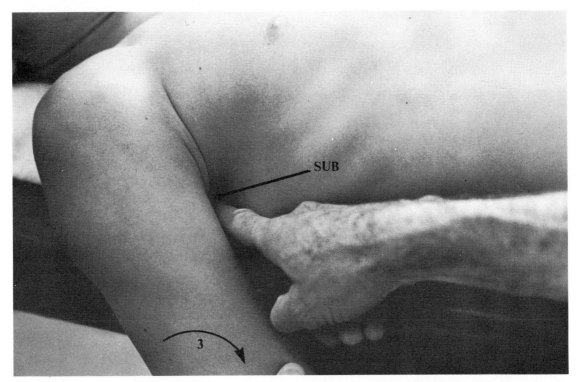

Fig. 81. **Treatment for a tender point in the subscapularis muscle dysfunction (SUB).** This is similar to treatment for latissimus dorsi dysfunction. This time, however, the emphasis is more on backward bending of the arm and internal rotation of the humerus without traction. The arm is held at about the 8 o'clock position.

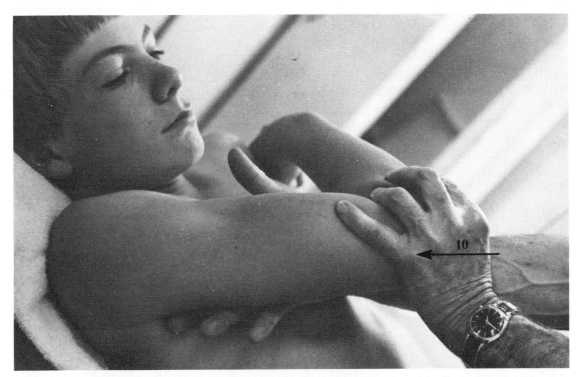

Fig. 82. **Treatment for adduction or "frozen" shoulder.** The patient hugs his elbow tightly to his chest and wishes that he could push it further into the chest. Hyperadduction can be achieved here by force from the elbow through the humerus. This raises the shoulder and rolls out the lower border of the scapula, giving much needed relief. The tender point is high on the lateral wall of the axilla but is seldom needed. Also, this procedure can be managed by the patient at home. He sits before a low table with his elbow on a pad on the table, leaning on it so as to force the shoulder up toward his ear.

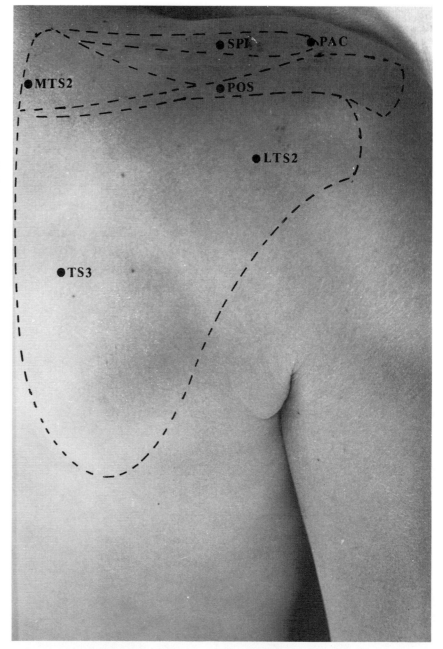

Fig. 83. **Posterior shoulder joint tender points.** At the top, behind the lateral head of the clavicle, is the posterior acromioclavicular tender point (PAC). Progressing medially (left) in the supraspinatus fossa is the tender point for the supraspinatus muscle (SPI). Next medially is the medial second thoracic shoulder joint tender point (MTS2). Laterally, two tender points, one on the spine of the scapula (POS) and the lateral second thoracic shoulder (LTS2) 2 cm. below are associated with intervertebral or rib dysfunctions. Five cm. beneath the spine of the scapula and 2 cm. lateral to the medial border is the third thoracic shoulder joint tender point (TS3), which is often associated with dysfunctions in the third segment. This last is by far the most common shoulder joint dysfunction.

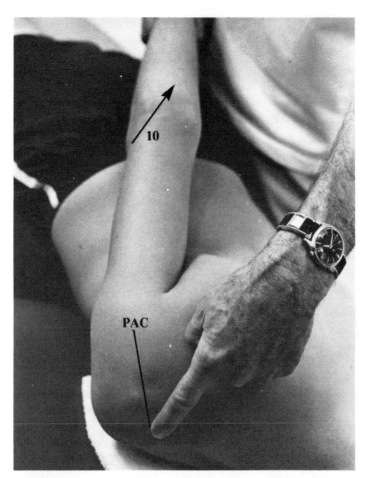

Fig. 84. **Treatment for posterior acromioclavicular dysfunction (PAC).** The physician stands at the left of the patient and pulls the right arm caudad, backward, and medialward with a force of 10 kg.

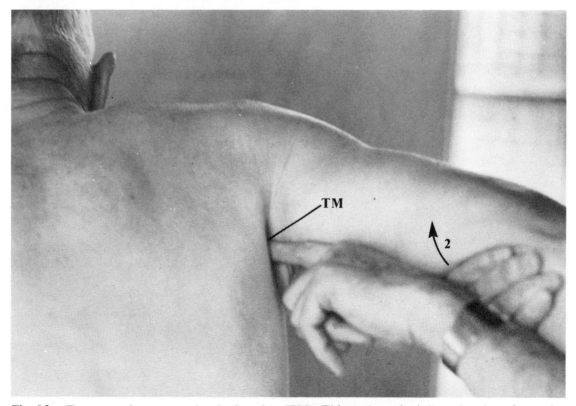

Fig. 85. **Treatment for teres major dysfunction (TM).** This tender point is lateral to the subscapular point in the posterior axilla. The fibers are shortened and relieved by moving the elbow backward and in, but markedly internally rotated.

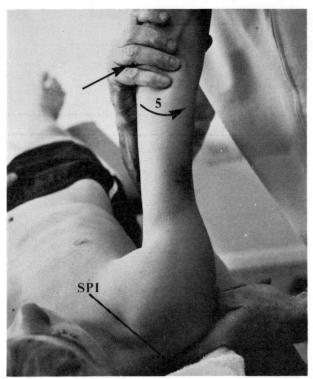

Fig. 86. **Treatment for supraspinatus muscle dysfunction.** The arm is raised about 45 degrees, abducted 45 degrees, and markedly externally rotated. This is a good site for the novice operator to practice feeling muscle changes. Even a well patient lying supine markedly relaxes the supraspinatus muscle in this position. The infraspinatus tender points are also good practice sites for developing palpation skills.

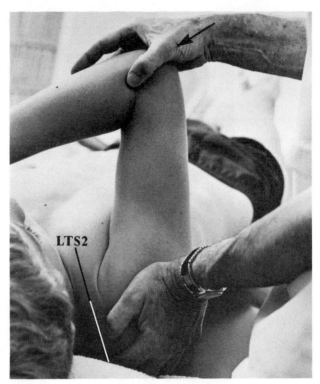

Fig. 87. **Treatment for the lateral second thoracic tender point on the infraspinatus muscle.** The patient lies supine, with the upper arm in the vertical position. The physician monitors the LTS2 tender point, and moderate abduction is introduced. Several shoulder dysfunctions are relieved similarly, with minor variations. The infraspinatus and the long head of the biceps are relieved by marked humeral flexion. Action for the upper fibers of the infraspinatus (lateral second thoracic shoulder) is less marked flexion with moderate abduction. Action for the short head of the biceps requires decreased flexion and moderate adduction.

Fig. 88. **Treatment for third thoracic tender point (TS3) on the infraspinatus muscle.** Maximal shortening of the lower fibers of the infraspinatus muscle is accomplished with this maneuver. Flexion is usually about 135 degrees.

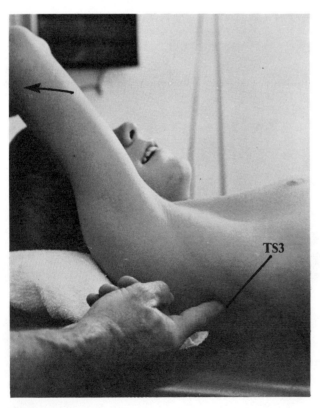

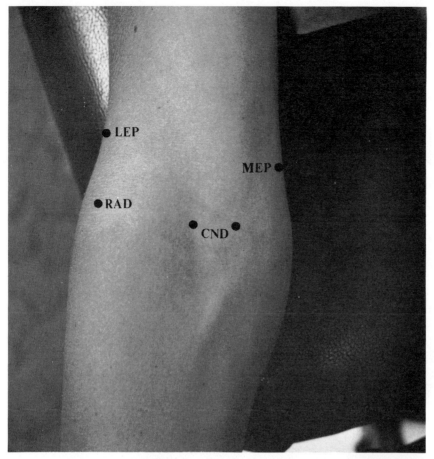

Fig. 89. **Anterior elbow tender points.** Tender points located high on the lateral epicondyle (LEP) usually result only from first thoracic or first rib disorders. Tender points high on the medial epicondyle (MEP) result from fourth thoracic or fourth rib disorders. The common epicondylitis discussed by orthopedic surgeons refers to a tender point on the tip of the lateral epicondyle. This is part of a tender point for a radial head dysfunction, by far the most common elbow dysfunction. The tender point is located on the anterolateral surface of the head of the radius (RAD). Two other tender points are located on either side of the coronoid process of the ulna (CND). They indicate flexion dysfunctions of the humeroulnar joint.

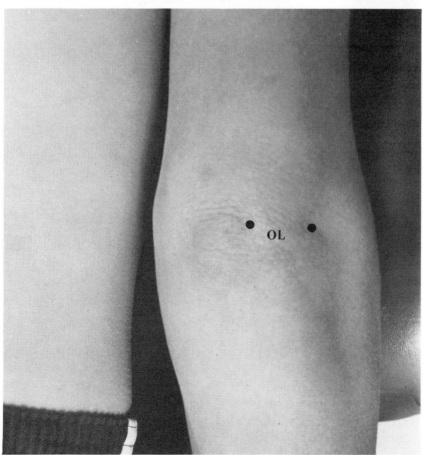

Fig. 90. **Posterior elbow tender points.** These are located on either side of the olecranon process of the ulna (OL) and indicate extension type dysfunctions of the elbow.

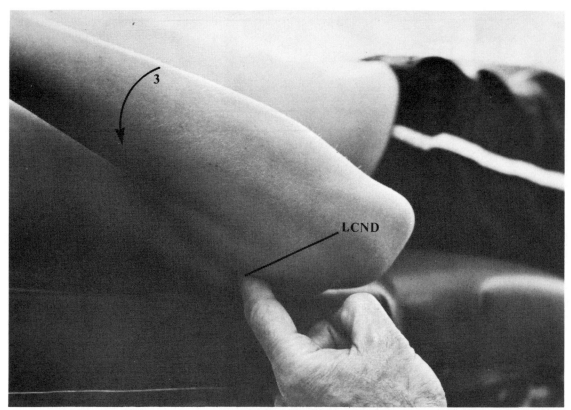

Fig. 91. **Treatment for lateral (LCND) or medial coronoid tender points.** These indicate a need for marked flexion of the elbow. The forearm is moderately pronated as well as flexed (palm forward). It is also pushed laterally as if to cause external rotation of the humerus.

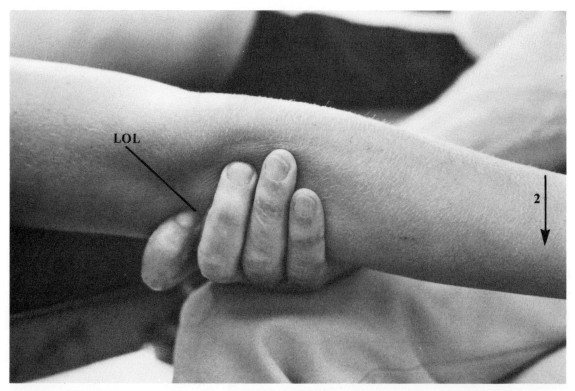

Fig. 92. **Treatment for lateral olecranon dysfunction (LOL).** The patient lies supine, with his elbow extended comfortably off the table and supported by the doctor's knee. The force is mild hyperextension, usually with supination and slight abduction (occasionally adduction). The index finger monitors tender point change. Treatment for medial olecranon dysfunction (MOL) is similar.

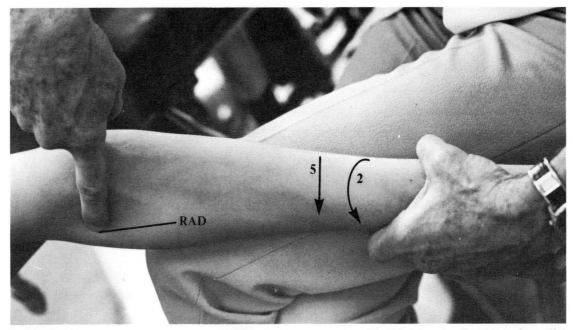

Fig. 93. (above) **Treatment for radial head dysfunction (RAD).** Although this elbow problem is treated in full extension, it is not forced. Supination and abduction are applied to the forearm in varying amounts of force; occasionally adduction is used.

Fig. 94. (left) **An uncommon radial head procedure (RAD).** The same tender point is used as in Figure 93. This time, however, the forearm is flexed and turned in so that the back of the hand touches the chest. Forces are pronation and internal rotation of the humerus.

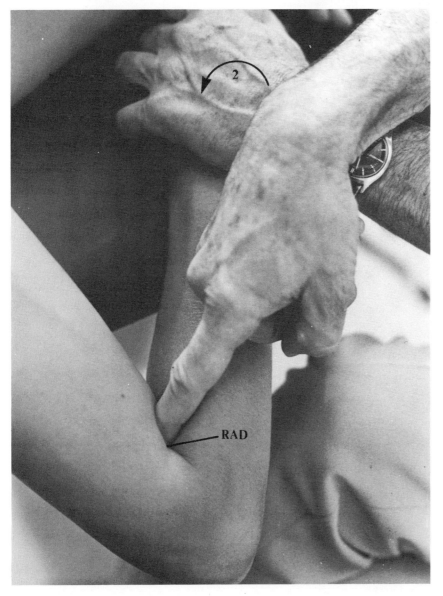

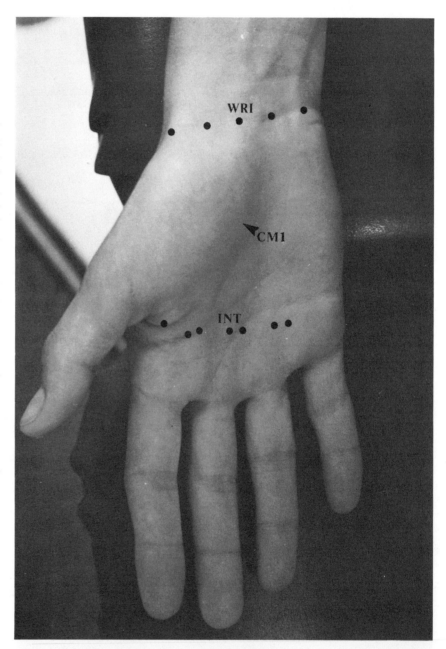

Fig. 95. **Wrist and hand tender points, palmar surface.** These wrist dysfunctions regularly respond to bending the hand over the tender point side (WRI). When maximum tissue release has been found by bending, it can be further improved by rotation. In this book the carpal joints are not discussed individually. The first carpometacarpal joint (CM1) is a tender point for the very common weak thumb, which is painful with use. Probing deeply on the proximal end of the first metacarpal bone from the hypothenar side of the palm will demonstrate this condition. The patient usually complains of pain at the distal end of the first metacarpal bone. A variation of this tender point is located on the dorsal or lateral surface of the proximal end of the first metacarpal bone. The interosseous joints (INT) are sites of the tender points for the other fingers. They are located on the palmar side of the hand, on the sides of the shafts of the metacarpal bones medially and laterally. Patients with these tender points usually complain of pain or "hanging up" of the first interphalangeal joint, so that they need to straighten the joint by pulling on the finger with the other hand.

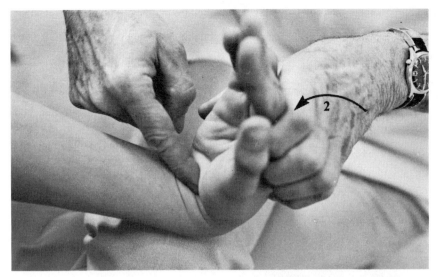

Fig. 96. **Treatment for palmar side wrist dysfunction.** The wrist is flexed markedly and possibly sidebent slightly over the tender point. Tissue changes are monitored while fine tuning is done by pronation or supination. This treatment will relieve many so-called carpal tunnel syndromes. For unexplained wrist pains, the proximal radial heads should be checked. (See Figures 89 and 93.)

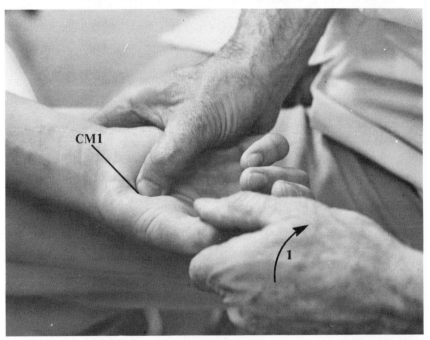

Fig. 97. **Treatment for first carpo-metacarpal joint dysfunction (CM1).** This is responsible for most lame thumbs. Deeply probed in the tough palmar tissue from the hypothenar side, it is relieved by marked rotation of the thumb toward the palm. Fine tuning is accomplished usually by a combination of abduction and flexion. Pain complained of is at one joint further down distally (the metacarpophalangeal).

Fig. 98. **Alternate method of treating the thumb.** The wrist is flexed as if to relieve flexed wrist dysfunction and pressure is applied toward the elbow at the distal head of the first metacarpal bone. This tender point is usually the variation mentioned in Figure 95.

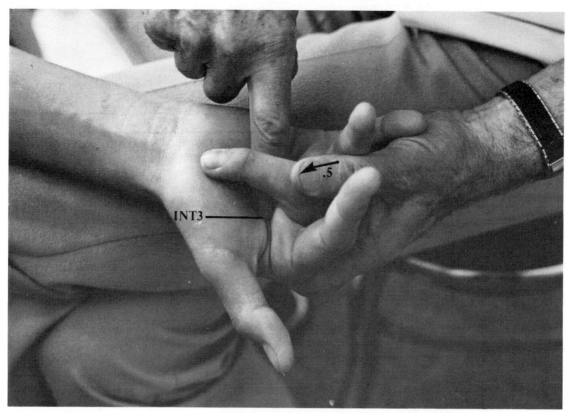

Fig. 99. **Treatment for third interosseous joint dysfunction (INT3).** This is typical of all metacarpophalangeal joint dysfunctions. One common patient complaint is a proximal interphalangeal joint that catches in flexion and cannot extend voluntarily. The metacarpophalangeal joint is stretched in flexion by the doctor. Minor adjustments are done by side pressure on the finger toward the affected palmar interosseous muscle. The patient thinks that his trouble is in the proximal interphalangeal joint.

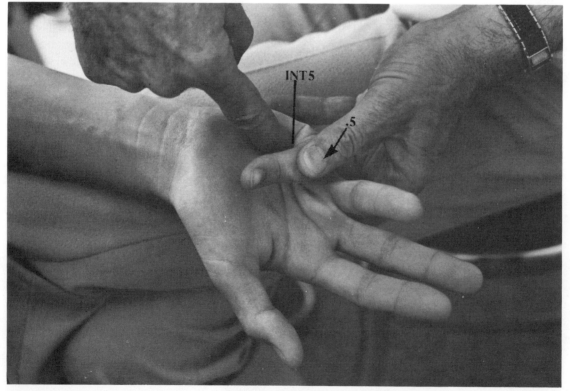

Fig. 100. **Fifth interosseous tender point (INT5).** Treatment is the same as that discussed in Figure 99.

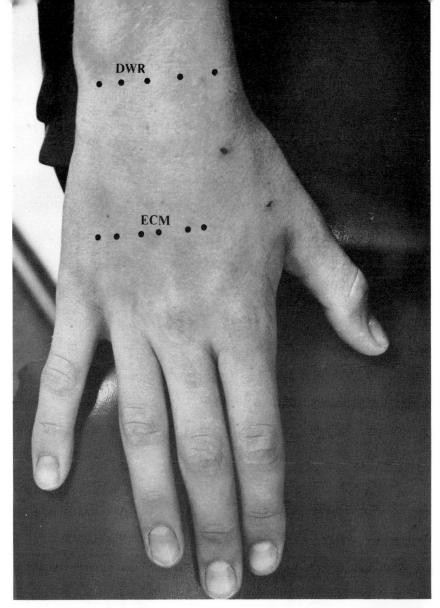

Fig. 101. (left) **Dorsal wrist tender points (DWR).** These are relieved similarly to the palmar wrist joint dysfunctions by folding the hand over the tender point. Minor adjustments are made by pronation or supination, and, perhaps, sidebending over the tender points. Extension carpometacarpal dysfunctions (ECM) are relatively rare disorders. They cause a vague ache in the hand on powerful flexion, and the patient may complain of weakness in his grip. They are relieved in extension somewhat like the interosseous tender points that are released in flexion.

Fig. 102. (right) **Treatment for dorsal wrist dysfunctions.** These respond to an extending of the hand over the tender point. Minor adjustments are done with pronation or supination and sidebending toward the side of the tender point. If there is unexplained wrist pain, check the proximal radial head. (See Figures 89 and 93.)

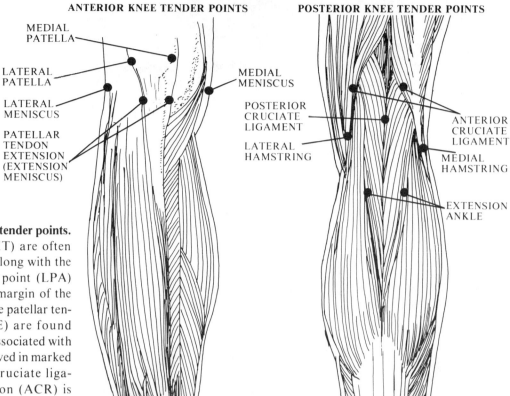

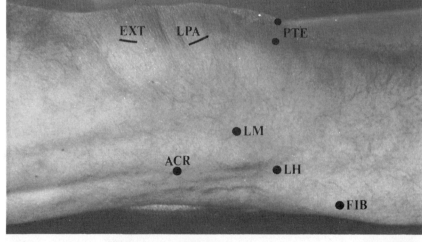

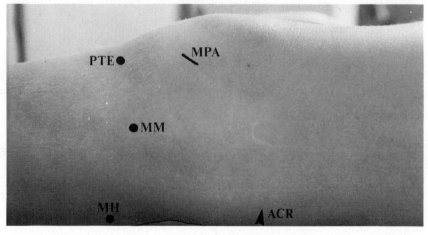

Fig. 103. **Lateral knee tender points.** Extensor muscles (EXT) are often tender and are found along with the lateral patellar tender point (LPA) on the lateral inferior margin of the patella. At the site of the patellar tendon attachment (PTE) are found tender points that are associated with a knee dysfunction relieved in marked extension. Anterior cruciate ligament reflex dysfunction (ACR) is found on the lateral hamstring in the popliteal area. The lateral hamstring muscle attachment (LH) is the site for a flexion abduction dysfunction. The posterior surface for the fibular head (FIB) is usually thought of as a knee problem, but it is nearly always associated with an ankle joint dysfunction and will be relieved by the correct stretch there. The lateral meniscus dysfunction tender point (LM) is located over the lateral meniscus and is less common than medial meniscus dysfunction.

Fig. 104. **Medial knee tender points.** The medial side of the patellar tendon (PTE) often presents a tender point associated with extension-type knee problems. On the medial inferior border of the patella (MPA) is the other knee disorder relieved by simple side pressure on the opposite side of the patella. A very common and important tender point for lame knee is located on the medial meniscus (MM). The knee will not fully extend. Another anterior cruciate tender point (ACR) is found on the medial hamstring muscle; it is also common. MH is at the attachment of the medial hamstring tendon.

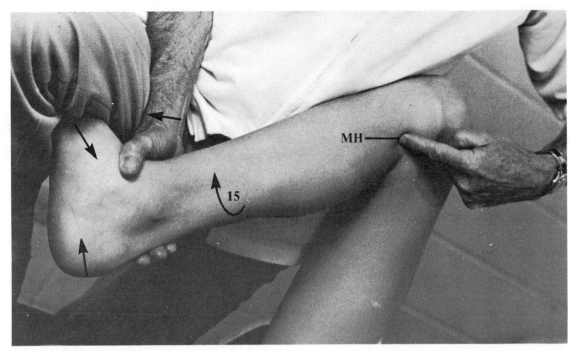

Fig. 105. **Treatment for medial hamstring dysfunction (MH).** This is a common knee joint dysfunction with a tender point at the attachment of the medial hamstring muscles. The patient lies supine and the knee is flexed to 60 degrees. The principal force applied is external rotation of the tibia on the femur. Note that the physician's foot is on the table, and that the front of the patient's foot is caught behind the doctor's knee. With his forearm resting on his own thigh, the doctor produces external rotation by forward and upward pressure under the ankle, causing 15 kg. of force to be exerted. The patient's foot is moved medialward.

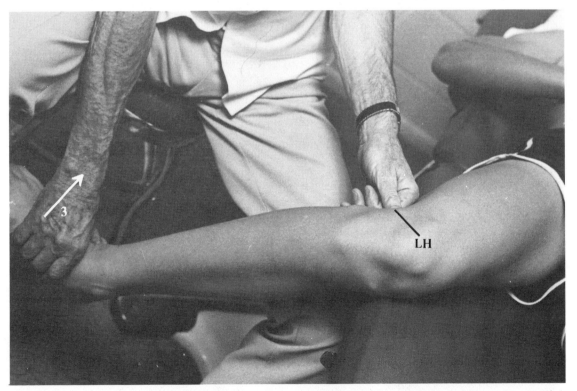

Fig. 106. **Treatment for lateral hamstring dysfunction (LH).** This disorder is less common than medial hamstring dysfunction. The patient lies supine with his leg abducted off the table. The physician sits beside the table and holds the patient's foot to cause external rotation, slight abduction, and about 30 degrees of flexion.

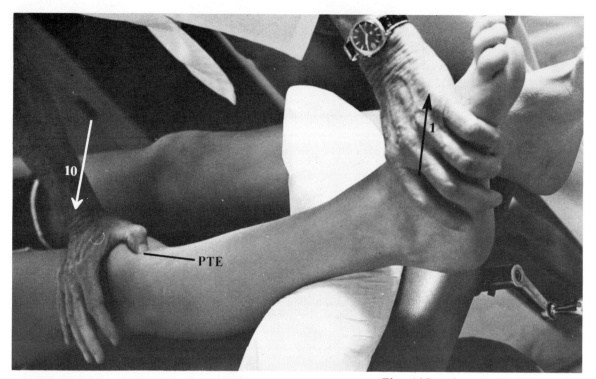

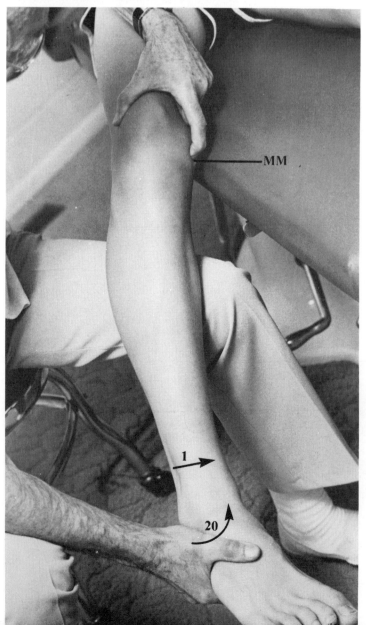

Fig. 107. (above) **Treatment for patellar tendon tender point (PTE).** This is an extension dysfunction. The knee is comfortable in full extension and is treated in hyperextension. Considerable force, usually about 10 kg., may be needed; one patient needed 25 kg. of force. Remember that the long lever you use makes you powerful, and the patient hasn't much strength to resist a hyperextended knee. The foot is rotated internally. The patient lies supine, with his lower leg resting on a pad placed just above the ankle. The physician applies pressure on the thigh just above the knee.

Fig. 108. (left) **Treatment for medial meniscus dysfunction (MM).** This very common and important knee problem can usually be relieved by internal rotation and slight adduction. The patient lies supine and abducts his thighs and lowers his leg off the table. With the knee flexed about 40 degrees, the doctor applies strong rotatory force internally and slight adduction of the knee over the edge of the table.

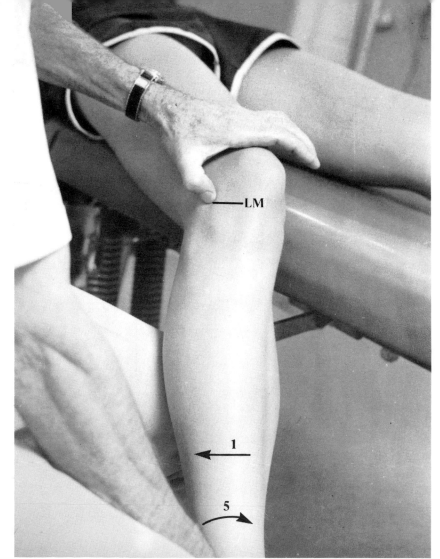

Fig. 109. (right) **Treatment for lateral meniscus dysfunction (LM).** This much less common knee problem begins in the same way and may in some cases be treated entirely the same as medial meniscus dysfunction, but the knee is often slightly abducted and externally rotated for release.

Fig. 110. (below) **Treatment for a tender point on the medial patella (MPA).** The physician's thumb presses on the lateral border of the patella with a force of about 2 kg. This simple procedure helps some knee problems surprisingly well. Likewise, a lateral patella tender point is relieved by pressure over the medial border of the patella.

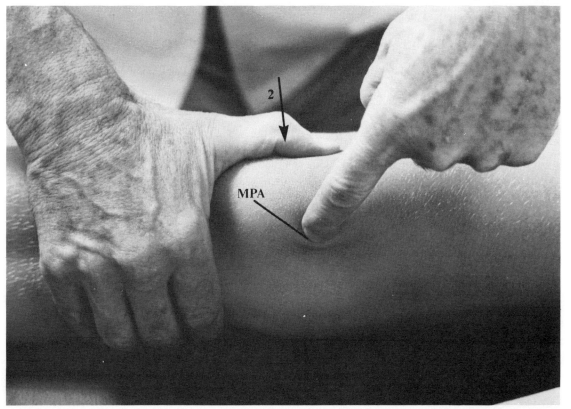

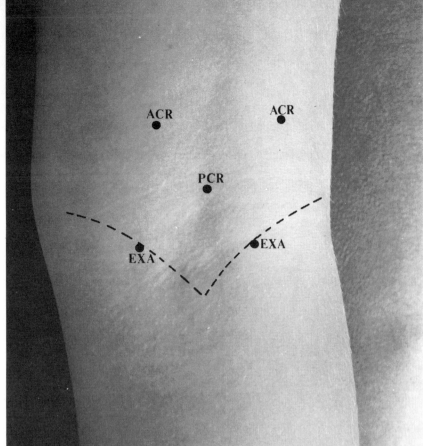

Fig. 111. **Posterior knee tender points.** ACR indicates the location of tender points for reflexes giving trouble from the anterior cruciate ligament. They are found in either hamstring muscle, or both, in the popliteal area. PCR indicates the location of the tender point for reflex dysfunction in the posterior cruciate ligament. It is found in the middle of the popliteal space. EXA indicates the location of tender points on the gastrocnemius muscle on the lower margins of the popliteal space. These are not actually knee problems, though they are often thought to be. They are released by hyperextension of the ankle (pointing the toes).

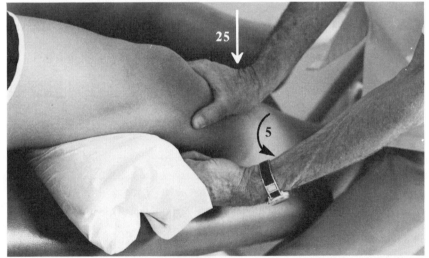

Fig. 112. **Treatment for anterior cruciate ligament dysfunction.** Because this is a shearing force and does not make use of long levers, this treatment requires much force (up to 25 kg.) The intention is to shorten the anterior cruciate ligament. The lower end of the femur is supported on the table with a rolled pillow, permitting a shearing force posteriorly to be applied over the proximal end of the tibia. Also, this hand causes internal rotation of the tibia. Note that the doctor's left hand monitors tissue changes in the hamstring muscle.

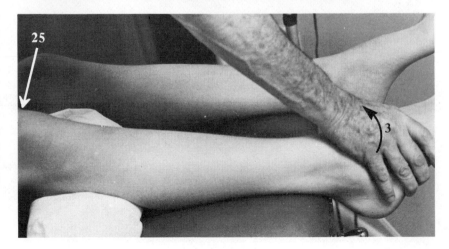

Fig. 113. **Treatment for posterior cruciate ligament dysfunction.** This is just the reverse of the treatment for anterior cruciate ligament problems. The pillow is placed under the proximal end of the tibia. The shearing force this time pushes the femur posteriorly in relation to the tibia, with the intention of shortening the reflexly dysfunctioning posterior cruciate ligament. Internal rotation of the tibia is the same.

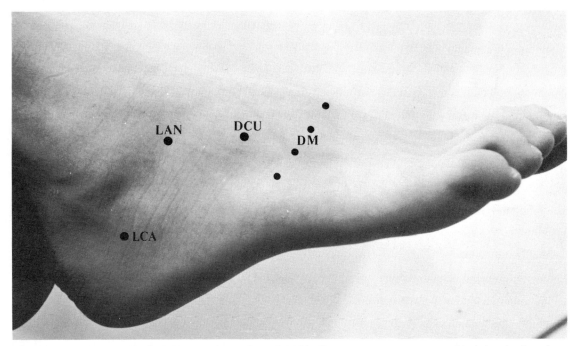

Fig. 114. **Lateral ankle and foot tender points.** LAN indicates a very common ankle joint dysfunction. It is the usual source of the so-called weak ankle or trick ankle. The tender point is located in a depression 2 cm. anterior and caudad from the lateral malleolus. The dorsal cuboid tender point (DCU) is 2 cm. further forward and down from the lateral ankle in a straight line. Dorsal metatarsal tender points (DM) indicate extension problems of the foot. LCA indicates the location of the tender point for a calcaneus reflexly held laterally; it is found 3 cm. caudad and posterior to the lateral malleolus.

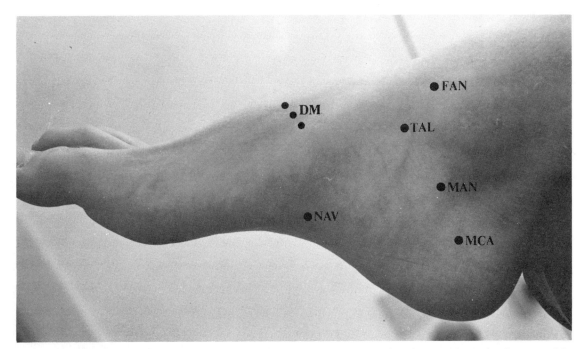

Fig. 115. **Medial tender points for the ankle and foot.** At the top right is a tender point for an ankle released in hyperflexion (FAN). It is located anteriorly and level with the medial malleolus, deep in the front of the ankle just medial to the tendon of the extensor digitorum longus. TAL indicates the location of a tender point on the tip of the talus. The medial ankle (MAN) locates a tender point for ankle dysfunction that is relieved in inversion. The medial calcaneus (MCA) locates a tender point for calcaneal joint dysfunction that is relieved similarly. There is also a tender point for joint dysfunction found around the tender navicular bone (NAV). Tender points for relatively uncommon metatarsal joint dysfunctions (DM) are relieved in foot extension (toes pointed toward the knee).

DORSAL FOOT TENDER POINTS

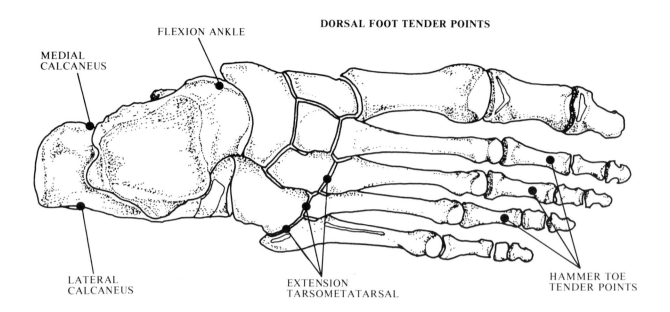

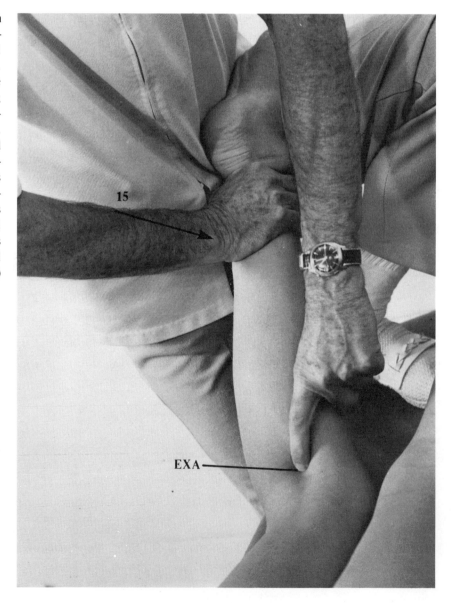

Fig. 116. **Treatment for extension ankle dysfunction (EXA).** Hyperextension of the ankle is accomplished with the patient in the prone position. The doctor stands at the foot of the table and places his foot on the table; the patient's foot rests over the doctor's thigh close to the ankle. Although the foot is a powerful mechanism, it is weak in this direction and may be injured if the force is applied too far down on the metatarsal bones. The force of the doctor's right hand is about 15 kg. cephalad and forward. The left hand monitors the tender point at the knee. (Helped by mild internal rotation of the tibia.)

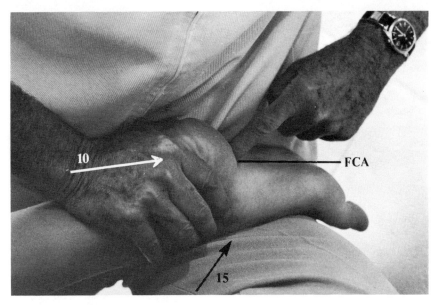

Fig. 117. **Treatment for flexion calcaneus dysfunction (FCA).** This treatment is discussed here because of its similarity to extension ankle treatment. The tender point is in the plantar surface of the foot at the anterior end of the calcaneus. The thigh pull is similar to that for extension ankle, but the physician's hand now forces the calcaneus into flexion with the foot. (This requires much less effort than the method shown in the article "Foot treatment without hand trauma."[12])

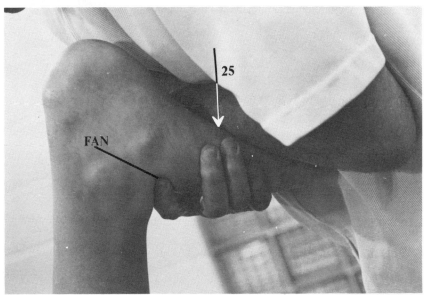

Fig. 118. **Treatment for flexion ankle dysfunction (FAN).** The patient lies prone with his knee flexed. The doctor stands at the foot of the table and leans heavily on his forearm, which is placed across the middle of the plantar surface of the foot. This causes marked flexion of the ankle (toes pointed toward the knee). The forearm is angled to produce straight ankle flexion, avoiding excessive pressure on the cuboid bone. The tender point is even with the medial malleolus and just medial to the extensor digitorum longus.

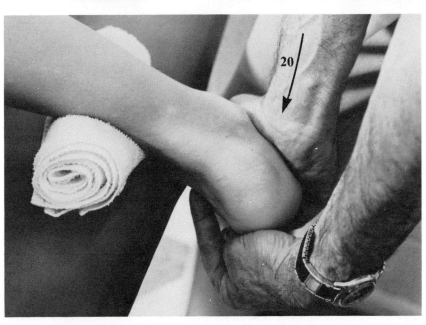

Fig. 119. **Treatment for medial ankle dysfunction.** This is a fairly common ankle problem, especially in people who wear out the lateral side of their shoes and break down their lateral longitudinal arches. The tender point is located 2 cm. below the medial malleolus in an arc about 2 cm. long. The patient lies on the left side, with his right knee bent 90 degrees and his foot suspended off the table. The ankle is padded from the table with a rolled-up towel. Force of up to 20 kg. is applied from the lateral side and causes hyperinversion. Note the left index finger monitoring the tender point. It helps to flex the other knee over the right knee to hold it down.

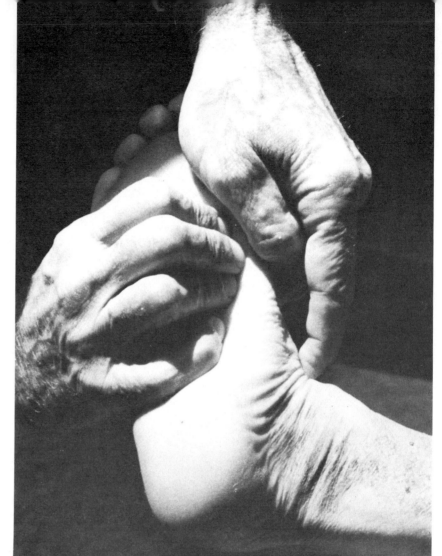

Fig. 120. **Treatment for talus dysfunction.** This ankle problem is slightly different from that of the medial ankle. The tender point is 2 to 3 cm. anterior to those for the medial ankle, in a depression on the anteromedial ankle where deep probing reaches the tender talus bone. Treatment is in fairly marked inversion, with 2 kg. internal rotation of the foot. The patient lies prone with his foot up.

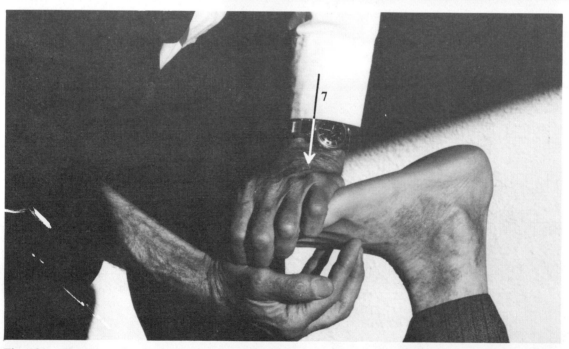

Fig. 121. **Treatment for dorsal metatarsal dysfunction.** These disorders appear to be much less common than plantar problems, but occasionally they are very important. The patient lies prone with his foot up. The doctor applies a fairly strong force on the ball of the foot to hyperextend the metatarsals. This problem is usually secondary to tarsal joint dysfunctions or ankle dysfunctions.

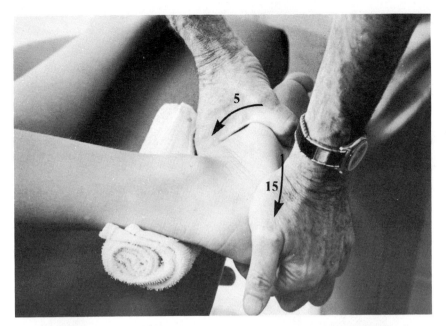

Fig. 122. **Treatment for medial calcaneus dysfunction.** The tender point for this problem is 3 cm. caudad and posterior to the medial malleolus. It is found with a heel that is held inverted in relation to the rest of the foot. A tenderness or even a spur under the lateral part of the tubercle of the calcaneus may have developed. At least in early stages these can be caused to regress. This dysfunction is corrected in exaggeration, as with medial ankle dysfunction. Here, a force of about 15 kg. is applied medialward, while counter-rotation of about 5 kg. of force is applied to the front of the foot. Also note the positioning of the hand over the calcaneus itself. Note the doctor's finger on the medial calcaneus monitoring the results of treatment.

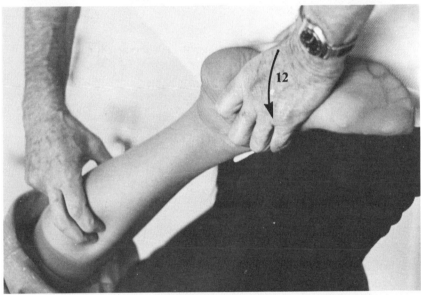

Fig. 123. **Treatment for flexion medial calcaneus dysfunction.** The tender point for the dysfunction is just posterior to the tibia on the medial margin of the calf on the medial fibers of the soleus. It is relieved by marked ankle extension. The calcaneus is flexed on the foot as in regular flexion calcaneus disorders; in addition, the calcaneus is inverted in relation to the front part of the foot.

Fig. 124. **Treatment for lateral ankle dysfunction.** The patient lies on his right side with his ankle suspended over the edge of the table. The tender point in front of the lateral malleolus is monitored with the right index finger, while the ankle is forced into marked eversion from above. A heavy force of about 25 kg. on the medial surface of the foot is applied. The ankle is padded from the table by a rolled-up towel. (This much force applied for 90 seconds can be tiring to the operator. The author uses the device of putting his olecranon process into his abdomen to save his arm.)

Fig. 125. **Treatment for lateral calcaneus dysfunction.** This treatment is similar to that for lateral ankle dysfunction, with the same variations used for the medial side of the foot. Apply force to evert the calcaneus with the right hand; with the left hand counter-rotate the distal part of the foot. The tender point is caudad and behind the lateral malleolus. The tender spot on the tubercle of the calcaneus in this case will be on the medial side, if present, because the calcaneus carried into eversion strikes its medial side on the ground. Note that the finger on the lateral side of the calcaneus is monitoring tissue change.

Fig. 126. **Treatment for dorsal cuboid dysfunction (DCU).** The tender point is just twice the distance from the lateral malleolus as that for the lateral ankle, and the action is about opposite. The patient lies in the prone position. The physician puts the lateral side of the patient's foot into inversion while monitoring the tender point. This dysfunction should be looked for in any case of resistant metatarsal troubles.

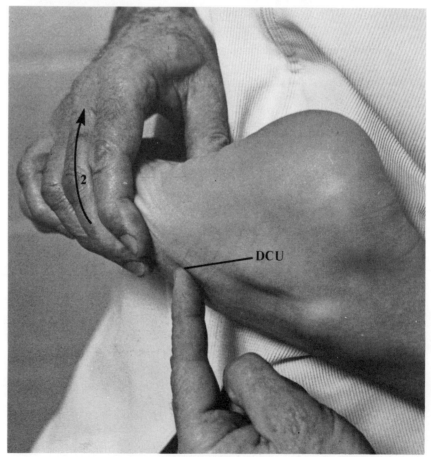

PLANTAR FOOT TENDER POINTS

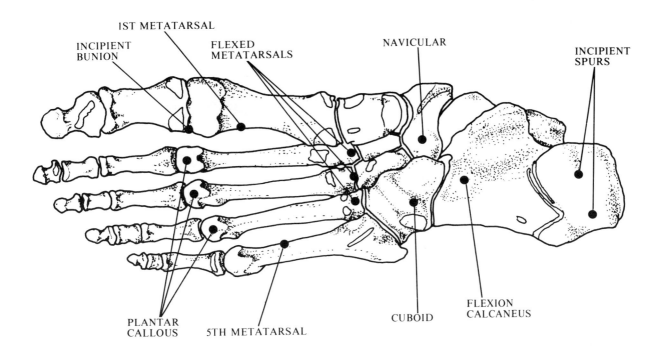

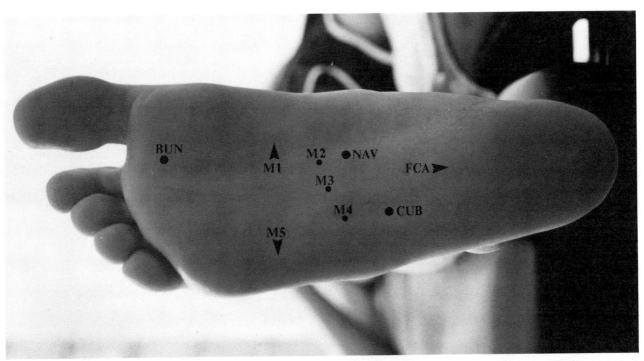

Fig. 127. **Plantar foot tender points.** NAV indicates the navicular tender point and FCA the tender point for flexed calcaneus dysfunction. The cuboid tender point (CUB) is found on the tubercle of the cuboid, which is deeply probed in the tough sole of the foot. It is 3 cm. posterior and medial to the prominence of the fifth metatarsal bone, which is used as a landmark. Flexion metatarsal tender points are located at the proximal ends, especially of the second, third, and fourth (M2, M3, and M4) metatarsals. The first and fifth metatarsal tender points (M1, M5) are found along the inner sides of the shafts. BUN indicates the location of the bunion tender point, which is associated with trouble on the medial side of the great toe. The bunion is so obvious that many physicians fail to also observe a very tender spot under the lateral sesamoid bone. Correction tends to abort or arrest an incipient bunion.

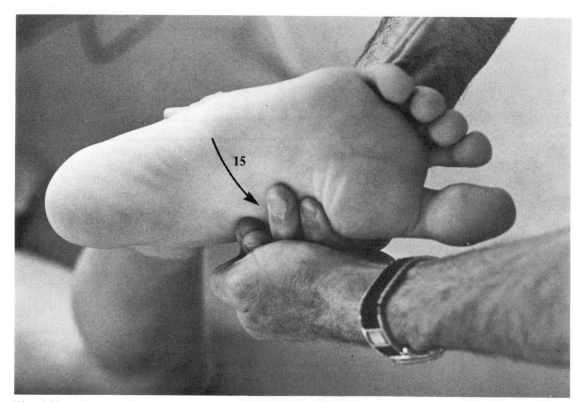

Fig. 128. (above) **Treatment for navicular dysfunction.** The patient lies prone. The physician wraps his index finger around the navicular bone, then reinforces it with his third finger and further reinforces that with fingers of the other hand. This device will provide enough force to cause rotation of the navicular bone into inversion. There is also slight flexion. The whole distal end of the foot may be allowed to rotate without appearing to cause any harm. The physician is too busy to monitor until after treatment.

Fig. 129. (right) **Treatment for plantar cuboid dysfunction (CUB).** The patient lies prone with his knee flexed. The doctor stands at the foot of the table. He grasps the lateral half of the metatarsals and presses cephalad with the distal end of his second metacarpal on the base of the patient's fifth metatarsal. The doctor uses a force of about 20 kg. This causes eversion and some extension of the cuboid in relation to the rest of the foot.

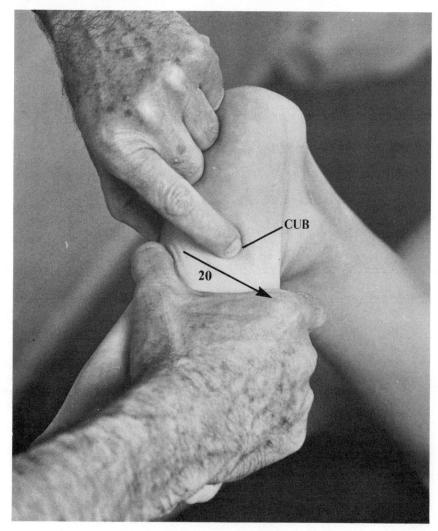

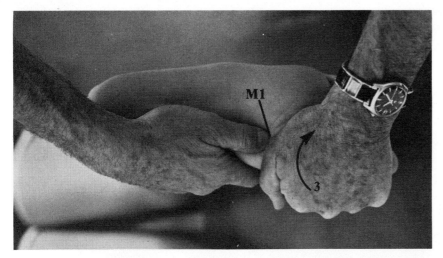

Fig. 130. **Treatment for first metatarsal dysfunction.** The tender point is found on the lateral inferior aspect of the shaft, indicating that a release may be achieved by rolling it into eversion. The force required is estimated at 3 kg.

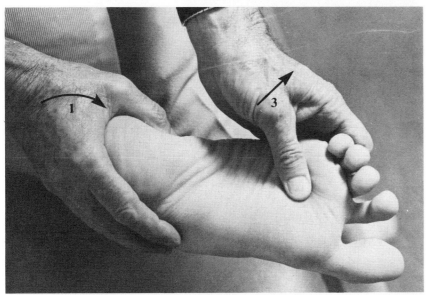

Fig. 131. **Treatment for flexion metatarsal dysfunction.** The tender points are monitored on the plantar surfaces of the metatarsals' proximal ends. The force required is estimated at 3 kg. The doctor pulls the second, third, and fourth metatarsals into flexion and external rotation (lengthwise, so as to point the toes laterally).

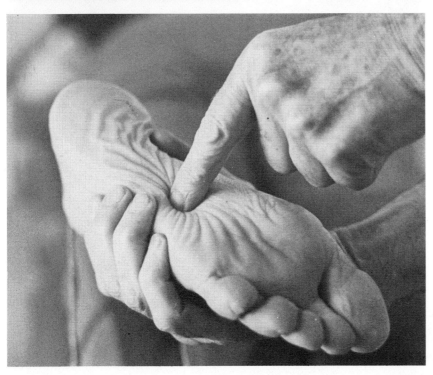

Fig. 132. **Treatment for fifth metatarsal dysfunction.** The tender point for this dysfunction is on the medial surface of the shaft of the fifth metatarsal. Treatment is just squeezing of the foot, which causes inversion of the fifth metatarsal bone.

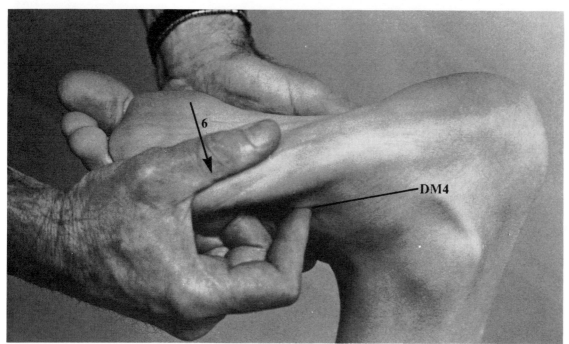

Fig. 133. (above) **Treatment for dorsal or extension fourth (DM4) or fifth metatarsal dysfunction.** This is a common type of "flat foot" disorder. Although many patients have a history of long and unsuccessful treatment, they may be relieved markedly and permanently. The patient lies prone and the doctor stands at the end of the table. He uses the base of his thumb to force the metatarsals into extension. The physician should not fail to search for concurrent medial ankle and cuboid problems. Most metatarsal problems have their sources in tarsal dysfunctions. Failing to correct these problems dooms the treatment, since they will recur.

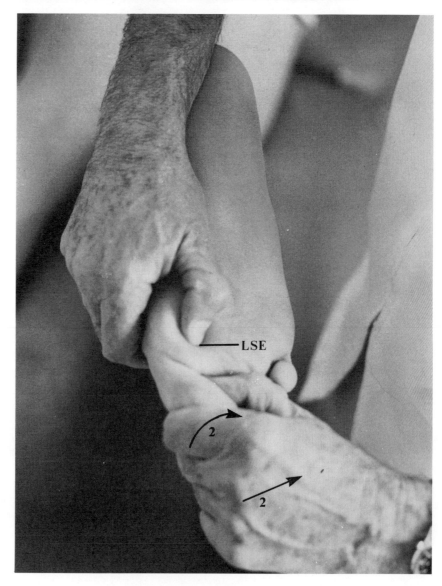

Fig. 134. (right) **Treatment for lateral sesamoid (LSE) or bunion disorders.** This joint can often be much improved in all but far advanced bunion problems. Success, to be lasting, must include the use of shoes that permit the great toe to function as normally as it can. These would be shoes with either a wide toe or a straight inner line. The lateral sesamoid bone is often very tender on the plantar side. The great toe is abducted (from the midline), everted, and flexed at the metatarsophalangeal joint.

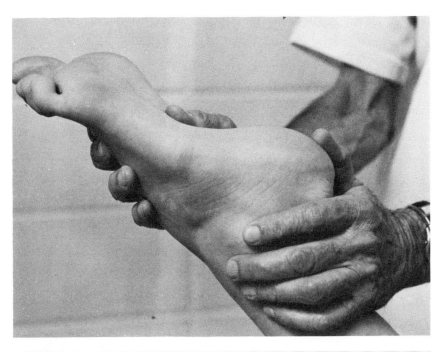

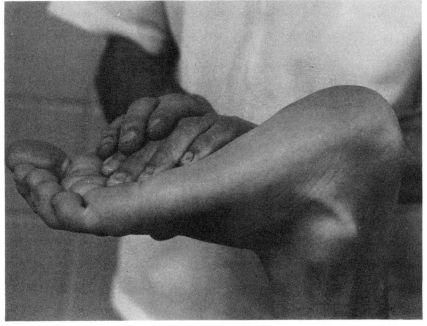

Figs. 135 and 136. **Flexion and extension foot dysfunctions and hammer toes.** A flexion foot is one with a high arch; in most cases the high arch is more commonly associated with reflex joint dysfunction. The reverse is the so-called flat foot, with a low arch. This also is associated with reflex joint dysfunction. Hammer toes are often treated locally. Attempts to stretch the extensor tendons of the toes have been fruitless. Like the metatarsals, their sources of trouble are located proximally, usually in the tarsal joints. Hammer toes can usually be helped if they are discovered before marked fibrosis has occurred. They are always associated with high arch foot troubles. The metatarsophalangeal extension of these problems is a direct result of foot hyperflexion. Treatment lies in the correction of the joint dysfunction that is causing the abnormal amount of tarsometatarsal flexion. These two photos show the marked changes in the metatarsophalangeal joints caused by flexion (Fig. 135) and extension (Fig. 136) of the foot.

Index to illustrations

Page	
24	Figs. 1-I to 1-III. Schematic representation of a joint and the role of its muscles.
26	Fig. 2. Modus operandi for ideal positioning for comfort.
41	Fig. 3. Tender points on the side of the head.
42	Fig. 4. Posterior cranial tender points.
43	Fig. 5. Treatment for occipitomastoid problems.
43	Fig. 6. Less common treatment for occipitomastoid disorders.
44	Fig. 7. Treatment for squamosal dysfunction.
44	Fig. 8. Treatment for posterior auricular dysfunction.
45	Fig. 9. Treatment for sphenoid disorders.
45	Fig. 10. Treatment for lateral canthus dysfunction.
46	Fig. 11. Treatment for coronal disorders.
46	Fig. 12. Treatment for infraorbital dysfunction.
47	Fig. 13. Facial, throat, and neck tender points.
48	Fig. 14. Treatment for supraorbital dysfunction.
48	Fig. 15. Treatment for nasal dysfunction.
48	Fig. 16. Treatment for masseter disorders.
49	Fig. 17. Bilateral compression.
49	Fig. 18. Treatment for lambdoidal dysfunction.
50	Fig. 19. Treatment for sphenobasilar torsion.
50	Fig. 20. Treatment for zygomatic dysfunction.
51	Fig. 21. Tender points for anterior neck disorders.
51	Fig. 22. Treatment and monitoring of the anterior first and second cervical joint dysfunctions.
52	Fig. 23. Treatment for anterior first and third cervical joint dysfunctions.
52	Fig. 24. Treatment for anterior fourth cervical joint dysfunction.
52	Fig. 25. Treatment for anterior fifth and sixth cervical joint dysfunctions.
53	Fig. 26. Treatment for anterior seventh cervical joint dysfunction.
53	Fig. 27. Treatment for anterior eighth cervical joint dysfunction.
53	Fig. 28. Treatment for lateral first cervical joint dysfunction.
54	Fig. 29. Tender points for posterior cervical joint dysfunctions.
54	Fig. 30. Treatment for posterior first and second cervical joint dysfunctions.
55	Fig. 31. Treatment for posterior fourth cervical joint dysfunction.
55	Fig. 32. Treatment for lower posterocervical and posterior upper thoracic joint dysfunctions.
56	Fig. 33. Tender spots for anterior dysfunctions in the thoracic area.
57	Figs. 34 and 35. Treatment for anterior first and second thoracic intervertebral joint disorders in forward bending.
57	Fig. 36. Treatment for anterior third and fourth thoracic joint dysfunctions.
58	Fig. 37. Treatment for anterior fifth and sixth thoracic joint dysfunctions.
58	Fig. 38. An alternate method of dealing with anterior fifth, sixth, seventh, and eighth intervertebral joint dysfunctions.
59	Fig. 39. Treatment for anterior dysfunctions of the seventh, eighth, and ninth thoracic areas.
59	Fig. 40. Closer view of Figure 39.
61	Fig. 41. Tender points of the posterior thorax.
61	Fig. 42. Prone position treatment of lower posterior cervical and upper posterior first and second thoracic joints.
62	Fig. 43. Similar treatment and monitoring of tender points for the posterior third, fourth, and fifth thoracic joints.
62	Fig. 44. Treatment for many posterior joint dysfunctions of the sixth thoracic through the second lumbar joints.
62	Fig. 45. Treatment for posterior mid- and lower thoracic dysfunctions with tender points in the midline on or near the spinous processes.
63	Fig. 46. Treatment for elevated first rib.
63	Fig. 47. Treatment for elevated ribs.
64	Fig. 48. Treatment for depressed first and second ribs.
64	Fig. 49. Treatment for depressed third to sixth ribs.
65	Fig. 50. Tender points for anterior lower thoracic and upper lumbar intervertebral joint dysfunctions.
66	Fig. 51. Treatment for forward bending joint dysfunctions from the ninth thoracic through the first lumbar levels.
67	Fig. 52. Treatment for regular anterior second lumbar dysfunction.
67	Fig. 53. Treatment for anterior third and fourth lumbar intervertebral joint dysfunctions.
68	Fig. 54. Treatment for abdominal second lumbar dysfunction.
68	Fig. 55. Treatment for anterior fifth lumbar joint dysfunction.
69	Fig. 56. Treatment for low-ilium sacroiliac dysfunction.
69	Fig. 57. Treatment for low-ilium sacroiliac dysfunction with flare out.
70	Fig. 58. Tender points in the groin.
70	Fig. 59. Treatment for anterolateral trochanter dysfunction.
71	Fig. 60. Treatment for iliacus dysfunction.
71	Fig. 61. Treatment for inguinal ligament dysfunction.
71	Fig. 62. Treatment for adductor dysfunction.

Page		Page	
72	Fig. 63. Tender points for posterior thoracic and upper lumbar dysfunctions.		surface.
73	Fig. 64. Tender points found on the buttocks.	89	Fig. 96. Treatment for palmar side wrist dysfunction.
74	Fig. 65. Treatment for third, fourth, and fifth lumbar intervertebral joint dysfunctions.	89	Fig. 97. Treatment for first carpometacarpal joint dysfunction.
74	Fig. 66. Treatment for lower-pole fifth lumbar dysfunction.	89	Fig. 98. Alternate method of treating the thumb.
74	Fig. 67. Treatment for high-ilium sacroiliac dysfunction.	90	Fig. 99. Treatment for third interosseous joint dysfunction.
75	Fig. 68. Treatment for midpole sacroiliac dysfunction.	90	Fig. 100. Fifth interosseous tender point.
75	Fig. 69. Treatment for high flare-out sacroiliac dysfunction.	91	Fig. 101. Dorsal wrist tender points.
		91	Fig. 102. Treatment for dorsal wrist dysfunction.
76	Fig. 70. Treatment for posteromedial trochanter dysfunction.	92	Fig. 103. Lateral knee tender points.
		92	Fig. 104. Medial knee tender points.
76	Fig. 71. Treatment for posterolateral trochanter dysfunction.	93	Fig. 105. Treatment for medial hamstring dysfunction.
77	Fig. 72. Treatment for lateral trochanter dysfunction.	93	Fig. 106. Treatment for lateral hamstring dysfunction.
77	Fig. 73. Treatment for piriformis muscle dysfunction.	94	Fig. 107. Treatment for patellar tendon tender point.
78	Fig. 74. Tender points for anterior shoulder joint dysfunction.	94	Fig. 108. Treatment for medial meniscus dysfunction.
78	Fig. 75. Treatment for dysfunction of the anterior acromioclavicular joint.	95	Fig. 109. Treatment for lateral meniscus dysfunction.
79	Fig. 76. Alternate treatment for anterior acromioclavicular joint dysfunction.	95	Fig. 110. Treatment for a tender point on the medial patella.
79	Fig. 77. Treatment for dysfunction of the long head of the biceps muscle.	96	Fig. 111. Posterior knee tender points.
		96	Fig. 112. Treatment for anterior cruciate ligament dysfunction.
79	Fig. 78. Treatment for shoulder bursal dysfunction.	96	Fig. 113. Treatment for posterior cruciate ligament dysfunction.
80	Fig. 79. Two axillary shoulder joint tender points.		
80	Fig. 80. Treatment for latissimus dorsi dysfunction.	97	Fig. 114. Lateral ankle and foot tender points.
		97	Fig. 115. Medial tender points for the ankle and foot.
81	Fig. 81. Treatment for subscapularis muscle dysfunction.	98	Fig. 116. Treatment for extension ankle dysfunction.
81	Fig. 82. Treatment for adduction or "frozen" shoulder.	99	Fig. 117. Treatment for flexion calcaneus dysfunction.
82	Fig. 83. Posterior shoulder joint tender points.	99	Fig. 118. Treatment for flexion ankle dysfunction.
83	Fig. 84. Treatment for posterior acromioclavicular dysfunction.	99	Fig. 119. Treatment for medial ankle dysfunction.
		100	Fig. 120. Treatment for talus dysfunction.
83	Fig. 85. Treatment for teres major dysfunction.	100	Fig. 121. Treatment for dorsal metatarsal dysfunction.
84	Fig. 86. Treatment for supraspinatus muscle dysfunction.	101	Fig. 122. Treatment for medial calcaneus dysfunction.
84	Fig. 87. Treatment for second thoracic tender point on the infraspinatus muscle.	101	Fig. 123. Treatment for flexion medial calcaneus dysfunction.
84	Fig. 88. Treatment for third thoracic tender point on the infraspinatus muscle.	101	Fig. 124. Treatment for lateral ankle dysfunction.
85	Fig. 89. Anterior elbow tender points.	102	Fig. 125. Treatment for lateral calcaneus dysfunction.
85	Fig. 90. Posterior elbow tender points.		
86	Fig. 91. Treatment for lateral or medial coronoid tender points.	102	Fig. 126. Treatment for dorsal cuboid dysfunction.
		103	Fig. 127. Plantar foot tender points.
86	Fig. 92. Treatment for lateral olecranon dysfunction.	104	Fig. 128. Treatment for navicular dysfunction.
		104	Fig. 129. Treatment for plantar cuboid dysfunction.
87	Fig. 93. Treatment for radial head dysfunction.		
87	Fig. 94. An uncommon radial head procedure.	105	Fig. 130. Treatment for first metatarsal dysfunction.
88	Fig. 95. Wrist and hand tender points, palmar		

Page	
105	Fig. 131. Treatment for flexion metatarsal dysfunction.
105	Fig. 132. Treatment for fifth metatarsal dysfunction.
106	Fig. 133. Treatment for dorsal or extension fourth or fifth metatarsal dysfunction.
106	Fig. 134. Treatment for lateral sesamoid or bunion disorders.
107	Figs. 135 and 136. Flexion and extension foot dysfunctions and hammer toes.

Index to anatomical drawings

Page	
40	Locations of tender points.
41	Tender points on side of head.
42	Tender points on back of head.
47	Tender points on face.
60	Posterior tender points.
92	Anterior knee and posterior knee tender points.
98	Dorsal foot tender points.
103	Plantar foot tender points.

References

1. Korr, I.M.: Proprioceptors and somatic dysfunction. JAOA 74:638-50, Mar 75

2. Northup, G.W.: Osteopathic medicine. An American reformation. American Osteopathic Association, Chicago, 1966, p. 16

3. Sutherland, W.G.: The cranial bowl. JAOA 43:348-53, Apr 44

4. Hoover, H.V.: Functional technic. Yearbook, Academy of Applied Osteopathy, Carmel, California, 1958, pp. 47-51

5. Jones, L.H.: Spontaneous release by positioning. D.O. 4:109-16, Jan 64

6. Ruddy, T.J.: Osteopathic rhythmic resistive duction therapy. Yearbook, Academy of Applied Osteopathy, Carmel, California, 1961, pp. 58-68

7. Mitchell, F.L.: Personal communication

8. Dorland's Illustrated medical dictionary, Ed. 24. W.B. Saunders, Philadelphia, 1965, p. 1478

9. Stein, J., Ed.: Random House Unabridged dictionary. Ed. 1. Random House, New York, 1966, p. 1403

10. Travell, J.: Basis of the multiple uses of local block of somatic trigger areas. Procaine infiltration and ethyl chloride spray. Mississippi Valley Med J 71:13-21, Jan 49

11. Owens, C.: An endocrine interpretation of Chapman's reflexes. Ed. 2. Chattanooga Printing and Engraving Company, Chattanooga, Tennessee, 1937

12. Jones, L.H.: Foot treatment without hand trauma. JAOA 72:481-9, Jan 73

Jones, L.H.: Missed anterior spinal lesions. A preliminary report. D.O. 6:75-9, Mar 66

Jones, L.H.: Strain and counterstrain. Rationale of manipulation. Address given at Fourth Annual Postgraduate Seminar of the American Academy of Osteopathy, Colorado Springs, May 25-27, 1972

Rumney, I.C.: Structural diagnosis and manipulative therapy. J Osteopathy 70:21-33, Jan 63. Revised version. D.O. 4:135-42, Sep 63